THE ROAD TO STOCKHOLM

István Hargittai is Professor of Chemistry at the Budapest University of
Technology and Economics, and Research Professor of the Hungarian
Academy of Sciences at Eötvös University. He is a member of the
Hungarian Academy of Sciences, foreign member of the Norwegian
Academy of Science and Letters, and member of the Academia Europaea.
He holds a Ph.D. degree from Eötvös University, D.Sc. degree from the
Hungarian Academy of Sciences, is Dr. h.c. of Moscow University, and
D.Sc. h.c. of the University of North Carolina. He has lectured in some
30 countries and taught as a visiting professor at several universities in
the USA. He has published extensively on structural chemistry and on
symmetry-related topics. His books include the Candid Science series
of his collected interviews with famous scientists. In December 2001,
Professor Hargittai gave an invited lecture on the subject of this book at
the Royal Swedish Academy of Sciences in Stockholm.

The Road to Stockholm

Nobel Prizes, Science, and Scientists

István Hargittai

Budapest University of Technology and Economics
Eötvös University
Hungarian Academy of Sciences

OXFORD
UNIVERSITY PRESS

OXFORD

Great Clarendon Street, Oxford OX2 6DP
Oxford University Press is a department of the University of Oxford.
It furthers the University's objective of excellence in research, scholarship,
and education by publishing worldwide in
Oxford New York
Auckland Bangkok Buenos Aires Cape Town Chennai
Dar es Salaam Delhi Hong Kong Istanbul Karachi Kolkata
Kuala Lumpur Madrid Melbourne Mexico City Mumbai Nairobi
São Paulo Shanghai Taipei Tokyo Toronto

Oxford is a registered trade mark of Oxford University Press
in the UK and in certain other countries

Published in the United States
by Oxford University Press Inc., New York

First published 2002
First published as an Oxford University Press paperback 2003

Library of Congress Cataloging in Publication Data
(Data available)

ISBN 0-19-860785-7 (Pbk)

1 3 5 7 9 10 8 6 4 2

Printed in Great Britain by
Clays Ltd., St. Ives plc

For Magdi

Foreword

The Nobel Prize is revered today almost as much as 50 years ago. Its several Nobel-bearing faculty enhanced greatly the esprit of the University of Chicago when I was one of its undergraduates. More important to me, however, was the 1946 awarding of the Nobel Prize in Physiology or Medicine to the 56-year old American geneticist, Hermann J. Muller. His presence was the reason why I applied to Indiana University for fall 1947 entry to their graduate school. This Southern Indiana school was still in the intellectual backwater, and I would have thought myself down marketing if Muller had not just moved to Indiana. Though I had been told that Indiana also then possessed several young genetics hot-shots, their names themselves would not have led to my going there.

As soon as I arrived in Bloomington, however, I discovered that Salvador Luria and Tracy Sonneborn were much more to my liking. They were much younger and did their research on the genetics of microorganisms, then genetics at its most exciting. In contrast, Muller's research still focused on *Drosophila*, the organism that he used for his Nobel Prize-winning 1926 demonstration that X-rays induce mutations. Twenty years later, work on the tiny fruit fly had a tired, almost arcane, feeling. In contrast, Muller's course on Advanced Genetics, which I took my first term at Indiana, was an eye-opener. It made by itself my coming to Indiana worthwhile. No one else had lived through so many great moments of genetics, and I eagerly went to every one of his lectures knowing I would learn more how knowledge about heredity unfolded. In going to Indiana, I wanted to get close to the true essence of the gene. Over the next three years, I greatly benefited from the thoughts of an individual who had come to this objective some 35 years earlier.

Seeing Muller in action was equally important in letting me learn first hand that Nobel Prize bearers are not super humans. He was very much the seasoned academic, more at ease with the past than with how to move into the future. But I, having no past, had no choice but to gamble my future on a path that did not yet exist. To make my mark, I had to settle on a big objective and stick with it until either I or someone else got to the top. Here my Indiana experience gave me the factual knowledge I needed. I had gone there with no interest in DNA, but by the time I left to seek my fortune in Europe, I wanted to think about nothing else. Being then obsessive in no way guaranteed my way to the Nobel Pantheon. But it sure helped!!

James D. Watson

19 November 2001

Preface

Ivan Gorchev, the unemployed sailor, was not yet 21 years old when he won the Nobel Prize in Physics. This is how a mystery novel first published in 1940 began.[1] The author knew that his broad readership would be familiar enough with the Nobel Prize to be duly impressed by Gorchev's accomplishment, and it was of secondary importance that he won the prize in a card game from an aged scientist who was on his way home from Stockholm.

The Nobel Prizes have had enormous worldwide fame and prestige. They have been awarded since 1901 and the year 2001 marked their centennial. The prizes are in the world news every October when the winners are announced, and every December when the awards are made, but little is known about the selection process and about the criteria according to which the Nobel Prizes are awarded. Even less is known about the scientific achievements for which the awards are made. Although the Nobel laureates are often in the limelight, they are almost as likely to be asked about their expectations of next year's fashion as about the work that led to their prize. Nonetheless, every October and December science gets into the limelight, too, and for a fleeting moment some selected scientists acquire celebrity status like that of sport champions and TV personalities. There are evening news reports daily in every country, with the world news being followed by the local news, and that in turn by the sports news. Imagine an extra segment every night on the latest scientific achievements. Well, for at least a couple of nights in October and December this does happen thanks to the Nobel Prize.

From early on the Nobel Prize has been the only science prize recognized by the wider public, and today it enjoys an improbably high prestige. It has a certain aura, and in an age when science and scientists are sometimes viewed with distrust, if not disrespect, the fascination with the Nobel Prize has hardly decreased. The Nobel Prize is overrated by the public, but this is not an unwelcome exaggeration. Science needs icons, since it usually suffers from an image of being impersonal.

This book grew out of a lecture, 'How to Win a Nobel Prize?', which I recently gave in Cambridge, sponsored jointly by the Laboratory of Molecular Biology and Peterhouse, the oldest college of the university. I did not coin the title and I did not know the answer to the question it posed, but I had talked with 70 Nobel laureates and I could share my impressions of their careers. My discussion here does not restrict itself to these personal encounters, but this is a personal account rather than a historical or sociological study. I am not a historian, neither am I a sociologist; I am a scientist. My book illuminates some decisive moments, both of the scientists' careers en route to the Nobel Prize and the science itself that is behind some of the contemporary Nobel Prizes. It also covers some discoverers for whom the ultimate recognition from Stockholm never materialized.

Conducting interviews with famous scientists has been a second education for me, and an excellent one at that, because I could learn about various fields from some of the best minds. As for myself, I am a PhD chemist; more specifically, a structural chemist. Structural chemistry is a subdivision of physical chemistry. In addition to my research, I have taught in several countries and lectured in more. From the beginning of my career, physics was closer to my chemistry than biology. At one point I even taught physics at the University of Connecticut for two semesters. As I started conducting these interviews, I gradually warmed to biology, and realized the closeness of all these disciplines to each other and, especially, the interwoven relationship between chemistry and the biomedical sciences. This was one of my lessons from these encounters.

The very first interview I recorded was with Nikolai Semenov (C56),[2] in 1965, but the bulk of my interviews were done during the late 1990s. The earliest Nobel Prize of my interviewees was Glen Seaborg's (C51), while the latest were Ahmed Zewail's (C99) and Günter Blobel's (M99). Four of my subjects who had not won the prize at the time of the interview have since become Nobel laureates.[3]

As the book uses a great many personal recollections, perhaps a caveat is in order. Different persons may view or remember the same events in different ways. James Watson remarked in the Preface of *The Double Helix*: 'I am aware that the other participants in this story would tell parts of it in other ways, sometimes because their memory of what happened differs from mine and, perhaps in even more cases, because no two people ever see the same events in exactly the same light.'[4] When Leo Szilard embarked on compiling some materials about the history of the Manhattan Project, he said 'that he was going to write down the facts. Not for publication, just for the information of God. When his colleague remarked that God might know the facts, Szilard replied that this might be so, but not "*this* version of the facts." '5

The book is written for the general reader, including scientists. Even the most learned professor of any given field is an outsider in everything else. For the interested non-scientist, taking a look at some turning points in the careers of outstanding scientists and into some of the most out-standing scientific discoveries of the second half of the twentieth century may be the most interesting aspect of this book. For the scientist, in addi-tion to having a look into other fields, perhaps some lesser-known intric-acies surrounding the Nobel Prize might be the main attraction. To all, my attempts to present and reflect on the human element in science and in the quest for its fruits, both as discoveries and awards go, may be of interest.

For a long time before I began this book, I used to have the same awe as everybody else about the Nobel Prize. As I have become acquainted with details of the careers of Nobel laureates and some of the intricacies of their winning the prize, my occasional disillusionment has been amply compensated for by the exciting information gathered about science and scientists. I did not have many preconceptions when I started working on this book, except some feelings of sadness for those who had been close to receiving the Nobel Prize and never did. Choosing one person, or a group of maximum three people, may sometimes lead to unjust decisions, for the Nobel Prize is a watershed. There is a higher probability of most of those who are left out disappearing into oblivion than is the case for the Nobel laureates, who will be listed in tabulations and mentioned in all encyclo-pedias. The Nobel Prize is held in so much higher esteem than any other prize in science that the contrast between the haves and have-nots is sharp. But is the institution of the Nobel Prize unique in producing such a contrast? No, it is not. In our age of fast communication and brief evening news, such contrasts are being produced daily in every facet of the human endeavor. Of course, those who would like to be better informed can go to the print media and eventually to books and the library. The same is true for science and the evaluation of scientific contributions. Scholars of science history, possessing also the benefit of hindsight, will have a detailed and more just overview of the discoveries and the personal shares in them. I venture to say that the institution of the Nobel Prize brought a certain technique of contrast and simplification into the sciences rather early, and the rest of human activities only followed after some delay.

Science has become an industry. Today we don't just complain about the low level of scientific literacy of the general public. Our worry is that even graduate students are not well versed in the history of their fields, that they are ignorant of the contributions of their famous predecessors. Arthur Kornberg (M59)[6] noted that American children are made to learn the names of their presidents, as British and French children have to learn about their respective kings, queens, and presidents, but much less about the great scientists. The same is true for the rest of the world. Our students, our children, the general public, all of us would benefit from knowing a little more about science and how it comes about because so much in our modern life depends on it. *I. H.*

Budapest, 2001

❦❦❦

Less than two years have passed since I completed the manuscript of the hardcover edition and less than one year from its publication. The present paperback edition is essentially the same as the hardcover edition except for the correction of a few (less than half a dozen) errors. The list of Nobel laureates in the sciences has been updated through 2002 and the list of Further reading has been augmented by two entries, both of 2001.

I would like to give here only a few comments concerning the science Nobel Prizes of the past three years. Each of the three science prizes in these years was awarded jointly to three scientists; thus the maximum possible number of people received the prizes as stipulated by the *Statutes*. Four Japanese scientists received Nobel Prizes in these three years, three in chemistry and one in physics. Some of the prizes, especially in chemistry, heavily favored discoveries that have resulted in applications, such as information and communication technology (Physics 2000), conductive polymers (Chemistry 2000), chirally catalyzed reactions (Chemistry 2001), and analytical techniques of biological macromolecules (Chemistry 2002). Three persons received prizes whose recognition was especially overdue, viz., Raymond Davis and Masatoshi Koshiba (Physics 2002) and Sydney Brenner (Physiology or Medicine 2002). The 2000 prize in Physiology or Medicine was an umbrella award in which three loosely related areas were combined into one shared prize. There was no doubt that the chemistry prize for conductive polymers went to the right people. However, the name of the Korean scientist, Dr. Hyung Chick Pyon, should also be recorded because he was the one who, while working in Shirakawa's laboratory, came upon the accidental discovery of a conducting polymer.[7]

In December 2001, my wife and I attended the centennial celebrations of the Nobel Prize in Stockholm. The events once again underlined the importance and international recognition of this institution. They culminated, as usual, on Alfred Nobel's birthday, December 10, in the award ceremony and the banquet that followed. On December 12, I gave an invited talk on the topic of the present book to the Royal Swedish Academy of Sciences with the title, 'For many are called, but few are chosen: The Road to Stockholm.' The lively discussion that followed the presentation and was then resumed after the dinner party following the lecture, showed how deeply the Swedish scientists are interested and involved in the Nobel Prize and how keen they are to tap the international reactions to their decisions.

During the time that has passed since the completion of the manuscript of the original hardcover edition of this book, I have not stopped being interested in the Nobel Prizes. In particular, I have been fascinated by the circumstances that lead to this highest recognition for some scientific discoveries and fail to lead to it for others. I have continued recording conversations with famous scientists and these in-depth interviews appear in the book series entitled *Candid Science*.[8]

Of course, it is impossible to give a recipe for winning a Nobel Prize, but Sydney Brenner made an attempt in his Banquet Speech on December 10 in Stockholm:

> First you must choose the right place to work with generous
> sponsors to support you. Cambridge and the Medical Research
> Council will do. Then you need to discover the right animal to
> work on—a worm such as C. elegans for example. Next, choose
> excellent colleagues who are willing to join you in the hard work
> you will need to do. How about John Sulston and Robert Horvitz
> [his co-winners] for a starter. You must also make sure that they
> can find other colleagues and students. Everybody will have to
> work hard. Finally, and most important of all, you must select a
> Nobel Committee which is enlightened and appreciative and has an
> excellent chairman with unquestioned discernment.[9]

Brenner's words reflect not only his sense of humor but also his magnanimity as his name was missing for a long time from the roster of Nobel laureates. In his case, the dilemma for the awarders may have been not so much whether he should receive the Nobel Prize or not but to choose the discovery for which he should be awarded. Once it was decided which particular discovery of Brenner would be awarded, selecting his co-winners must have been straightforward.

During the past 10 months or so, the hardcover edition of *The Road to Stockholm* has been reviewed extensively and very favorably. I am grateful for those reviews. The book seems to achieve its goal in bringing science and scientists as well as the Nobel Prize into human proximity. I appreciate the decision of Oxford University Press to bring out this paperback edition and thus make the book more widely available.

Budapest, January 2003   István Hargittai

❦❦❦

A word about language. English is not my first language but I love it and feel that I can communicate my thoughts in English. In my mother tongue, Hungarian, there is no gender and perhaps this has conditioned me to be more sensitive to what we call political correctness than people who grow up with gender in their language. It bothers me when I use 'he' rather than 'he or she'. On the other hand, it would appear forced to keep employing the 'he or she' routine. I felt I had to make this note here lest I appear to be further strengthening the male-dominated character of Nobel dealings.

List of Plates

1 The Nobel Prize and Sweden

1. The Nobel Medal of the physiology or medicine prize (© *The Nobel Foundation; courtesy of Ingmar Bergström, Stockholm*)
2. Announcing the Nobel Prize in Chemistry for the discovery of fullerenes at the Royal Swedish Academy of Sciences on 10 October 1996 (*Photo by I. Hargittai*)
3. Nobel Prize ceremony in 1978 (*Photo by Svenskt Pressfoto, courtesy of Edit Ernster, Stockholm*)

2 The Nobel Prize and national politics

1. British Nobel laureates at the British Embassy in Stockholm in 1991 during the 90th anniversary of the first awarding of the Nobel Prize (*Courtesy of John Vane, London*)
2. Richard Kuhn (*Courtesy of the Oesper Collection and William Jensen, University of Cincinnati*)
3. Adolf Butenandt (*Courtesy of the Oesper Collection and William Jensen, University of Cincinnati*)

3 Who wins Nobel Prizes?

1. Kary Mullis in 2000 (*Photo by M. Hargittai*)
2. Linus Pauling (*Courtesy of the MRC LMB Archives, Cambridge*)
3. William Astbury (*Courtesy of Simon Phillips, Leeds*)
4. Murray Gell-Mann, Yuval Ne'eman, and Abdus Salam at the Eighth Nobel Symposium in Göteborg, 1968 (*Courtesy of Yuval Ne'eman, Tel Aviv*)
5. James Watson measuring a model of DNA in Cambridge, early 1950s (*Courtesy of the MRC LMB Archives, Cambridge*)
6. Erwin Chargaff in New York City, late 1940s (*Courtesy of Erwin Chargaff, New York City*)
7. Stanley Prusiner (*Courtesy of Stanley Prusiner, San Francisco*)
8. Charles Weissmann in London, 2000 (*Photo by I. Hargittai*)
9. The authors of the paper that reported the discovery of buckminsterfullerene, C_{60} (*Courtesy of Harold Kroto, Falmer, East Sussex*)
10. Maria Goeppert-Mayer in Basel, 1949 (*Photo by, and courtesy of Ingmar Bergström, Stockholm*)

4 Discoveries

1. Albert Einstein in 1921 (*Courtesy of the Weizmann Institute of Science, the Weizmann Archives, Rehovot, Israel*)

2. Max Planck (*Courtesy of the Oesper Collection and William Jensen, University of Cincinnati*)

3. Albert Szent-Györgyi and Dorothy Hodgkin in New York City in the late 1940s (*Courtesy of Andrew Szent-Györgyi, Waltham, Massachusetts*)

4. Tsung-Dao Lee and Chen Ning Yang visiting a physics laboratory in Stockholm in 1957 (*Photo by, and courtesy of Ingmar Bergström, Stockholm*)

5. Frederick Sanger at the time of his first Nobel Prize (*Courtesy of the MRC LMB Archives, Cambridge*)

6. Marshall Nirenberg in Bethesda, Maryland, 1999 (*Photo by I. Hargittai*)

5 Overcoming adversity

1. Rosalyn Yalow in 1948 (*Courtesy of Rosalyn Yalow, New York City*)

2. Henry Taube on the day of the announcement of his Nobel Prize (*Courtesy of Henry Taube, Stanford, California*)

3. Petr Kapitsa in Cambridge, late 1920s (*Photo by, and courtesy of David Shoenberg, Cambridge*)

4. Lev Landau in Moscow, 1937 (*Photo by, and courtesy of David Shoenberg, Cambridge*)

6 What turned you to science?

1. Eugene Wigner and the author in Austin, Texas, 1969 (*By an unknown photographer*)

2. James Black in London, 1998 (*Photo by I. Hargittai*)

3. Carleton Gajdusek with children in Tongariki Island in the New Hebrides (*Courtesy of Carleton Gajdusek, Amsterdam*)

7 Venue

1. Francis Crick and James Watson in Cambridge, about 1950 (*Courtesy of Cold Spring Harbor Laboratory Archives*)

2. Joseph Thomson and Ernest Rutherford outside the Royal Society Mond Laboratory in Cambridge, 1933 (*Photo by, and courtesy of, David Shoenberg, Cambridge*)

12 *Who did not win*

Contents

I

The Nobel Prize and Sweden

A Swedish professor, who is deeply interested in Nobel affairs but annoyed by the criticism about some omissions among the Nobel laureates, exclaimed recently: 'The Academy could get rid of all criticism by forgetting about the prizes. But then little Sweden in a dark corner of Northern Europe would lose the unique contact with the world events in science.'[1] For us outsiders it is hard to imagine how emotional the cold Swedes may become about the Nobel Prize. It is a national affair in Sweden. Come October even the cab drivers seem to participate in the guessing game. Of course, the Nobel Prize is the only science award, worldwide, that is appreciated by the general public, not just the scientists.

The Nobel Prize has five categories: physics, chemistry, physiology or medicine, literature, and peace. The Royal Swedish Academy of Sciences awards the physics and chemistry prizes, the Karolinska Institute, a medical research institute and graduate school in Stockholm, awards the physiology or medicine prize, the Swedish Academy (consisting of 18 Swedish writers) the literature prize, and a committee appointed by the Norwegian Parliament, the Storting, the peace prize. At the time of Nobel's *Will*, in 1895, Sweden and Norway were united under the Swedish Crown. However, relations were strained and the involvement of Norway was, in Swedish nationalistic eyes, a thorn in the side. Norway became independent in 1905, but its participation in Nobel affairs has remained unchanged.

There is a Nobel memorial prize in economic sciences awarded by the Swedish Bank, which is also administered by the Royal Swedish Academy of Sciences. The first Nobel Prizes were awarded in 1901, the first memorial prize in economics in 1969. Eventually, the Nobel Foundation decided that no more new prizes would be established. Although the foundation can overturn its own decisions, it seems unlikely that new prizes will be initiated. Our story concerns the three science prizes.

There are several disciplines missing among the prizes, of which mathematics and biology are the most conspicuous. Various hypotheses exist, ranging from the scholarly to the frivolous, as to why no mathematics prize exists. According to a popular one, a mathematician stole the affection of a woman in whom Nobel was also interested, but nobody knows for sure. For biology, the prize in physiology or medicine comes close, especially since experimental medicine has been greatly favored over clinical medicine by the medical Nobel committee.

During the first decades of the twentieth century, the physics and chemistry prizes overlapped on several occasions in topics such as radioactivity. With the spectacular progress in biochemistry, the chemistry and physiology or medicine prizes have frequently been close to each other. At some point it was estimated that about 40% of all the chemistry and physiology or medicine prizes went to biochemistry.[2] This means that there were more prizes in biochemistry than in either chemistry or in physiology or medicine. At Nobel's time this discipline hardly existed, but when it took off it became very strong in Sweden. There has been much less overlap between the physics and physiology or medicine prizes. Nonetheless, there have been cases when physicists were awarded the physiology or medicine prize.[3] Institutionalized co-ordination has existed between the physics and chemistry and the chemistry and physiology or medicine committees and, starting from 2000, even between the physics and physiology or medicine committees.[4]

The Nobel organization consists of the Nobel Foundation, the prize-awarding institutions, and the small Nobel committees elected by these institutions. The Nobel Foundation does not participate in the selection of the laureates. It is a private institution and the Swedish government has no influence over its activities. It takes care of the money, administers the investments, makes the money available for the prizes and for the selection process, and organizes and funds the ceremonies. The Nobel committees make the primary recommendations to the prize-awarding institutions, which are solely responsible for deciding on the prizes. When we speak about the 'Nobel Committee', it may mean only the committee in a particular subject because there is no overall Nobel Committee.

The Will

The relevant part of Alfred Nobel's (1833–96) *Will*, on whose basis the Nobel Foundation was established, reads as follows:

> The whole of my remaining realizable estate shall be dealt with in the following way: the capital, invested in safe securities by my executors, shall constitute a fund, the interest on which shall be annually distributed in the form of prizes to those who, during the preceding year, shall have conferred the greatest benefit on mankind. The said interest shall be divided into five equal parts, which shall be apportioned as follows: one part to the person who shall have made the most important discovery or invention within the field of physics; one part to the person who shall have made the most important chemical discovery or improvement; one part to the person who shall have made the most important discovery within the domain of physiology or medicine; one part to the person who shall have produced in the field of literature the most outstanding work in an ideal direction; and one part to the person who shall have done the most or the best work for fraternity between nations, for the abolition or reduction of standing armies and for the holding and promotion of peace congresses. The prizes for physics and chemistry shall be awarded by the Swedish Academy of Sciences; that for physiological or medical works by the Caroline Institute in Stockholm; that for literature by the Academy in Stockholm, and that for champions of peace by a committee of five persons to be elected by the Norwegian Storting. It is my express wish that in awarding the prizes no consideration whatever shall be given to the nationality of the candidates, but that the most worthy shall receive the prize, whether he be Scandinavian or not.
>
> *Paris, November 27, 1895 Alfred Bernhard Nobel*[5]

Today we take it for granted that the Nobel Prize is awarded internationally. However, when Nobel prepared his *Will* there was uproar in Sweden that he wanted to let foreigners pick up so much Swedish money. The Swedish king at the time, Oscar II, also protested. He wanted the prize to be given to Swedes only. The king even tried to get the Nobel family to change Nobel's *Will*, especially with reference to the peace prize. He feared that it would lead to 'controversies and diverse complications'. Here is an excerpt from the conversation between the king and Emanuel Nobel, the family representative:

The king: Your uncle has been influenced by peace fanatics, and
particularly by women.

Emanuel: Your Majesty perhaps agrees with General Moltke:
'Eternal peace—that is only a dream, and not a beautiful one
either.'

The king: Did he say that, 'A dream, and not a beautiful one
either?' [He repeated it twice.][6]

The opponents of Nobel's intentions found them unpatriotic. They did
not realize how much Sweden was to benefit from this prize in the long
run. Neither did they care that Nobel, a chemical engineer by training,
had earned his fortune in foreign countries. Nobel's fortune originated
mainly from explosives. After his death, his assistant, Ragnar Sohlman,[7]
shrewdly transferred his wealth from France to a Swedish farm Nobel
had established. There were also horses on that farm, and according to
a little-known French law a man's domicile was where he kept his horses.
Thus Nobel's monies were sent home legally, without taxes being paid
on them.

Oscar II did not attend the first prize-awarding ceremonies in 1901;
the crown prince substituted for him. The famous painting shows the
crown prince handing over the first ever Nobel Prize to Wilhelm Conrad
Röntgen (P01). Yet another blemish dimmed the luster of the first ceremonies. The full-size medals of the Nobel Prize were not ready and the
laureates received them later. The small replicas of the medals were
available, though, from the very beginning. These are given, as an incentive, to all the members of the Nobel committees and members of the
Academy and the Karolinska Institute who participate in the voting for
the Nobel Prize.

The *Statutes* (see later in this chapter) prescribe that the Nobel
laureates receive, in addition to the money, a diploma and a gold medal.
Erik Lindberg, the Swedish sculptor, designed the medals for the science
and literature prizes, often called the Swedish medals. They feature the
portrait of Alfred Nobel and the years of his birth and death in Latin,
NAT—MDCCCXXXIII OB—MDCCCXCVI. The reverse side bears
the inscription 'Inventas vitam juvat excoluisse per artes' and the symbol
of the respective prize-awarding institution. Each medal weighs about 200
grams and has a diameter of 66 millimeters. Up to 1980, they were made
of 23-carat gold; since then, they have been 18-carat gold, plated with 24-

carat gold. The medals are struck by Myntverket, the Swedish Mint, in Eskilstuna.

From 1902 on the king of Sweden attended the ceremonies and when he did not, it was not because he opposed the prize. In fact, the participation of the royal family of Sweden in the award has been one of the fundamental sources of the prize's high prestige. There is no other occasion when 'Princes assemble to pay their respects to molecules'.[8]

'Nobelitis' and 'Nobelomania'

The Nobel Prize has great benefits for Swedish science. It brings many of the best scientists, who would like the Swedish scientific community to learn about their latest results, to Sweden all the year round. The negative side effect is that the students who attend these seminars are liable to think that this is the routine everywhere. They hear the compliments about Swedish science, which may and may not be correct, and may be misled by them.[9]

There is a disease, called Nobelitis, that afflicts the person who is close, or thinks he is close, to getting the prize. Then his life centers on this possibility, making him miserable. This is why Jacques Monod (M65) said, 'The Prize is very good for science and very bad for the scientists.'

When Göran Liljestrand retired after having served for four decades as secretary of the Nobel Committee for Physiology or Medicine, he was asked about his experience. He saw no aspect of the prize that would not have both positive and negative effects. When a journalist pressed him hard to find something that has only positive effects, he said the prize forces the professors of the Karolinska Institute to read the scientific literature.[9] According to some, the Karolinska Institute has two kinds of professor, one who likes to give out Nobel Prizes and the other who likes to receive them. When somebody becomes too obsessive about giving out the prize, it is called Nobelomania.

Because of the high prestige of the Nobel Prize, the international community subjects the Nobel decisions to strict scrutiny. Conversely, most Nobel decisions successfully withstand this critical examination and this adds to its prestige. The institution of the Nobel Prize comes closest to what could be a world authority in science, an organization which, even if possible to set up, would never work.

The money

There is a lot of money attached to the Nobel Prize today. The amount of award money has grown with fluctuations and occasional jumps in value. Today's Nobel Prizes are approximately one and half times higher in value than a hundred years ago. A few examples in Swedish kroner (US$ 1 was approximately SEK 9 in 2000):

1901	150 800	1940	138 600	1980	880 000
1910	140 700	1950	164 300	1990	4 000 000
1920	134 100	1960	226 000	1995	7 200 000
1930	172 900	1970	400 000	2000	9 000 000

In most countries the Nobel Prize is tax-exempt, with the United States being a notable exception.

As of 31 December 1998, the market value of the invested capital of the Nobel Foundation was (in million SEK) 3162, its operating income 504, and operating expenses 23. It has been stipulated that the money from Nobel's estate can be used exclusively for the prizes and closely related activities. The latter include reimbursement of expenses for the members of the Nobel committees and outside experts. There are also small Nobel institutes, which were originally created to check the claims of various discoveries under consideration for the prize. Although it was soon realized that such checking would be impossible and superfluous, some of the Nobel institutes have continued as small research institutions supported by the Nobel Foundation. Expenses that are not closely related to the Nobel Prizes, like the production of Nobel posters depicting the new Nobel laureates and their achievements each year, or the Nobel e-Museum, although both being part of the Nobel institution, must be supported from outside sources.

One of the most conspicuous uses of the award money occurred when it had been allocated even before the prize had been won. Albert Einstein (P21) married his first wife, Mileva Marić, a former fellow student at the Swiss Federal Institute of Technology, in 1903. By then they had a one-year-old daughter. At the time of their daughter's birth Einstein was working at the Bern patent office and Mileva had returned to her native Serbia. The daughter is supposed to have died in an epidemic soon after her parents' wedding. The marriage ended in divorce in 1919, and by then it was widely anticipated that Einstein would sooner or later win the Nobel Prize. The divorce settlement stipulated that, whenever this should take place, the prize money would go to Marić. In 1922 Einstein was awarded the physics prize for 1921 and the prize money was about $32 000.[10]

Michael Smith's (C93) share of the prize was about half a million US dollars. He gave most of it to research on schizophrenia, science outreach programs, and the encouragement of women in science.[11] He kept enough money to invite his guests to the award ceremony in Stockholm. This is also what Dorothy Hodgkin (C64) did, and she entrusted the rest of the prize money to be used over the years in small amounts for various good causes.[12] As soon as Günter Blobel (M99) learned about his Nobel Prize in October 1999, he announced that he would donate the bulk of it to the reconstruction of Dresden, Germany. Philip Anderson (P77) built the prize money into his family home. He received one-third of the physics prize in 1977, whose total prize money that year was well under a hundred thousand dollars. The Swedish economy was in a depression and the finances of the Nobel Foundation were not handled in the best way.[13]

A dramatic allocation of the prize monies happened with the 1923 prize in physiology or medicine. Frederick Banting and John Macleod received the prize for discovering insulin, a work done primarily by Banting and his student assistant Charles Best. The prize-winning work appeared in 1922 under the authorship of Banting, Best, and Macleod. Banting shared his money with Best, and Macleod followed suit by sharing his money with J. B. Collip, who joined in the work at a later stage. This prize was one of the few in which the 'preceding year' stipulation of Nobel's *Will* was literally observed. However, the haste may have led to errors in the evaluation of who did what. Banting and Best were the principal discoverers. A formal explanation for Best's omission stated that no nomination had arrived on his behalf.[14]

Festivities

It is an exciting day in Stockholm, in every October around the 10th, when the physics and chemistry Nobel Prizes are announced, with a few hours between them, at a press conference in the Academy.[15] The Secretary of the Academy, the chair of the respective Nobel committee, and one member of the committee, the one most knowledgeable about the discovery, sit at a small table facing a large horseshoe-shaped table with many reporters behind it. Other visitors, mostly schoolchildren, sit along the walls of the beautiful room, thickly decorated with oil paintings of great scientists. The great educational value of the event is not overlooked, and once the prizewinners have been announced there is a lecture at the Academy for the whole membership, followed by a dinner. This dinner is sober, almost frugal, but the Academy members are relaxed, finally, after the great tension involved in reaching the decisions.

On one occasion Carl Djerassi was the lecturer. Djerassi is a world-famous scientist, best known for the first synthesis of an oral contraceptive, who might have received a chemistry Nobel Prize. Lately he has been an author and playwright, and he often deals with topics related to the Nobel Prize. Probably because of his closeness to the prize, which proved elusive for him, on the occasion of the dinner party at the Academy, he was presented with a 'Nobel Peas Prize'. This was a can of green peas wrapped in the Swedish flag. Djerassi[16] saw in this joke recognition of his literary activities, a proof that his Swedish friends had read his novel, *Cantor's Dilemma*.[17] In this book the fictitious Professor Cantor and his research associate receive the Nobel Prize for a brilliant hypothesis on tumorigenesis. Various details of the Nobel ceremonies are recounted in the book, based on Djerassi's painstaking studies of them as experienced by friends he had successfully nominated for the prize. He describes the main ceremonial dinner party, during which the queen makes comments on eating habits in different parts of the world. The Americans cut with the knife and fork, then put the knife down, switch the fork to the other hand, and use the fork only. Europeans use their knife and fork in a continuous fashion, which is much more efficient than the American way. When it comes to eating peas, the Americans chase the peas with their fork because by then they have deposited their knife. The Europeans' habits diverge. Continental people, continuing to use their knives and forks, shovel the peas onto their forks with their knives. The British, however, who also use both utensils simultaneously, keep the prongs of the fork pointing toward the plate, as when cutting a piece of meat. It is extremely difficult to gather and keep the peas on this inverted fork because they roll off. One way out is to smash the peas with your knife and thus immobilize them. It is an amusing story, told in the context of the Nobel dinner.

The exhausting program of the 'Nobel week' made a flamboyant laureate suggest that a royal messenger should deliver the Nobel Prize instead of having the ceremonies in Stockholm. Kary Mullis (C93) then elaborated,

> It was the most exhausting affair of my life for two weeks. I came back dead but I loved it. . . . The most charming episode of the whole time I was there happened as I was leaving. I was taking a little boat from Malmö to Copenhagen. As I was getting on that boat, there was no fanfare associated with it, but somebody recognized me on the boat. He had a big hat and a big feather on it, a weird looking Swede. He saw me coming on and he took off his hat and he bowed very deeply and he said, Dr. Mullis, the Swedish people love you. He said it real loud, and the people clapped, and I started crying.[18]

The new laureates and their families spend about two weeks in December in Sweden. A typical itinerary is summarized below. Variations are possible except for the award ceremony, which is always on 10 December, the day of Alfred Nobel's death.

6 December   The Nobel laureates arrive in Stockholm. The winner often trades in his first-class ticket for several coach tickets to pay for the transportation of family members. It happens that the airlines, especially SAS, upgrade them on the spot, anyway. Upon arrival, each Nobel laureate is assigned a car with driver and an attaché by the Swedish Foreign Ministry for the duration of their stay.

7 December   Press conferences, receptions by the prize-awarding bodies, and dinners with the Nobel committees are organized.

8 December   The Nobel laureate has one obligation, and that is to give a lecture on the occasion of the Nobel Prize. These presentations are usually given on this day, though in some cases the laureates return to Stockholm at a later date to give their lectures. The laureate has a six-month period in which to deliver the Nobel lecture. Of course, there is no censure if he does not. Röntgen came to Stockholm in 1901, and received the prize but did not give a lecture; he said he would return for it, but never did.

The Nobel lectures are collected in a volume for each year and published in a book series, *Les Prix Nobel, The Nobel Prizes*. Various scientific journals, and a book series that groups the different fields separately, also publish them. Obviously, there is no Röntgen contribution in the printed lectures and a few others are also missing. Murray Gell-Mann (P69), for example, gave his lecture but never submitted it for publication. On the other hand, Christiaan Eijkmaan (M29) did not deliver a lecture because of his illness, but submitted the text. Within a year of the award, Eijkman died.

In addition to their lectures the laureates participate in various celebrations. It is customary for the embassies of their countries to give them receptions on 9 December, and the chief rabbi of Stockholm invites the Jewish laureates for a special service and a reception.

10 December The anniversary of Alfred Nobel's death and presentation day. In the morning there is a rehearsal, and the actual ceremony is in the afternoon in the Stockholm Concert Hall. Great attention is given to every detail, utilizing the experience of a hundred years. The flowers decorating the concert hall come from the Italian town of San Remo, where Alfred Nobel spent many of his last years. The Stockholm Philharmonic Orchestra plays pieces of music associated with the countries of the laureates. The ceremony is introduced by the chairman of the Board of the Nobel Foundation, followed by a representative of each of the five awarding committees describing the research, discoveries, and achievements for which the respective Nobel Prizes are being awarded. The peace prize is presented in Oslo earlier in the day.

Much has been made of the order in which Alfred Nobel enumerated his prizes, starting with physics, followed by chemistry, physiology or medicine, and literature. The protocol of the Nobel ceremonies follows strict rules. If there is one physics laureate, his wife goes into the dining room on the arm of the king. Further rules regulate the assignment of the other laureates and their spouses to members of the royal family when going into the dining room, and as regards the seating order. The hierarchy of subjects determines also the order in which the laureates receive the diploma and the medal from the king during the prize-awarding ceremony. In 1963, one of the three physics laureates was a woman, Maria Goeppert-Mayer, and she would have been the first among the physicists according to the alphabetical order. This might have posed a problem for the protocol people. However, they chose Eugene Wigner to be the first among the physicists, hence no new precedence was established. Wigner's primacy could be justified on the grounds that he was given half the physics prize while Goeppert-Mayer and Hans Jensen shared the other half.

A formal dinner at the Stockholm City Hall follows the prize-giving ceremony. Following coffee, the laureates address the audience with a brief speech. Each subject is represented by one speech, so if there are two or three laureates, they select one representative from among themselves.

There are plenty of opportunities to let the students of Stockholm University and the Royal Swedish Institute of Technology see the laureates and talk with them. Besides, the press, including the TV companies, conduct interviews with the laureates, asking them the most relevant and irrelevant questions.

11 December   The laureates receive their checks at the Nobel Foundation. In the evening there is a dinner in the Royal Palace.

12 December   There are more meetings with students, more interviews with the press, and more cultural events.

13 December   This is the last day of the official program of the Nobel week and it is also Lucia day in Sweden. At 6 a.m. staff of the hotel, dressed in white and carrying candles, wake the Nobel laureates. They symbolize the queen of light and her entourage, who signify the return of light to the northern hemisphere.

Rita Levi-Montalcini (M86) made the most poetic reference to the Nobel festivities. She received the prize for her discovery of the nerve growth factor (NGF), which first revealed itself to her 'in the anticipatory, pre-Carnival atmosphere of Rio de Janeiro' in 1952. Then, in December 1986,

> NGF appeared in public under floodlights, amid the splendor
> of a vast hall adorned for celebration, in the presence of Royals
> of Sweden, of princes, of ladies in rich and gala dresses, and
> gentlemen in tuxedos. Wrapped in a black mantle, he bowed
> before the king and, for a moment, lowered the veil covering his
> face. We recognized each other in a matter of seconds when I saw
> him looking for me among the applauding crowd. He then replaced
> his veil and disappeared as suddenly as he had appeared.[19]

Statutes

For the first 100 years of the Nobel Prize, there have been 162 laureates in physics, 135 in chemistry, 172 in physiology or medicine, 97 in literature, and 107 in peace. The economics prize has been awarded since 1969, with a total of 46 awardees as of 2000.

Nobel's *Will* is remarkable in that in one page it covered so much ground, and relatively few additional rules have been needed to make the whole operation work for decades in a fast-changing world. *The Statutes of the Nobel Foundation*[20] were created to clarify and govern the Nobel arrangements in all matters that were not unambiguous enough in the *Will*. They were approved by the king of Sweden in 1900 and amended subsequently, most recently in 1995. I will quote only a few of these statutes that are important for the selection of the laureates.

The *Will* refers to works 'during the preceding year'. The *Statutes* makes it clear in §2 that older works can also be considered 'if their significance has not become apparent until recently'. This means there is no time limit in considering a work for the prize, although in the first decades this so-called recency rule was interpreted in a way that made it possible to only consider work carried out within the preceding two decades.

The rule causing the most controversies, in §4, refers to the number of people that can share a prize: 'A prize may be equally divided between two works, each of which may be considered to merit a prize. If two or three persons have produced a work, which is to be rewarded, the prize shall be awarded to them jointly. In no case may a prize be divided between more than three persons.' This three-person rule is often referred to as part of Nobel's *Will*, which it is not. It is the *Statutes* that stipulate this limitation. It is a big question whether the three-person rule could or should be changed, and how. While nothing is unchangeable in the *Statutes*, with over a hundred years of tradition the Nobel institution is rather conservative, and any change may involve a long procedure. There is a curious statement at the end of §4: 'Each prize-awarding body shall be competent to decide whether the prize it is entitled to award may be conferred upon an institution or association.' So far only peace prizes, awarded by the Norwegian Nobel Committee, have been bestowed on institutions or associations, like Amnesty International (1977), or to an individual and an organization, like Joseph Rotblat and the Pugwash Conferences on Science and World Affairs (1995). It is also according to §4 that there is no posthumous Nobel Prize.[21]

The existing practice of awarding the science prizes attributes everything to individuals, sometimes overstressing their roles. This is a sociological phenomenon. It is not only the Nobel Prize but the general perception that tends to attribute to individuals more than they actually did. According to Kenneth Wilson (P82), we all learned Newton's laws, but we learned methods for solving Newton's laws which are not due to Newton himself. Newton's laws are much more powerful today than they were originally because of the incredible amount of research that has been done on the techniques for solving these laws. Wilson's primary example is the motion of the moon, with which Newton had trouble in his own time. The first successful calculation that was necessary to solve the problem came about after Newton's death, around 1740. Until the problem with the moon had been cleared up, many people just paid no attention to Newton. Ironically, Wilson's prize should have probably been shared with two other physicists who had contributed significantly to the achievements in phase-transition studies, for which he, alone, was singled out (see, p. 241).

The two present possibilities for sharing the prize have been utilized from time to time. One of these is to divide the prize into two parts where the two parts have nothing to do with each other. This was the case, for example, when in 1978 Petr Kapitsa received half of the physics prize, with the other half going to Arno Penzias and Robert Wilson. This seldom happens. More typical is when the prize is divided between two or three persons for the same discovery.

As for possible changes, physics professor Anders Bárány,[4] long-time secretary of the Nobel Committee for Physics and science historian, would find it ideal for the science prizes to have an arrangement similar to that practiced by the Norwegians for the peace prize. The physics prize in 1984 was given to Carlo Rubbia and Simon van der Meer for the discovery of the W and Z bosons. However, the prize could have been divided, one half between Rubbia and van der Meer with the other half going to the European Nuclear Research Center, CERN,[22] in Geneva. Then CERN could have had the diploma on the wall saying that they received a Nobel Prize in physics. It would have been nice not only for CERN but for the Nobel Prize, too. Of course, not everybody agrees with Bárány, and the opposition insists on rewarding those 'delta function' contributions that stick out sharply against the background of the rest. However, the proposed change would still allow this, but would make it possible to reward the large groups at the same time.

Although §6 says that Swedish citizenship or membership of the adjudication body shall not be a necessary qualification for election to membership of a Nobel committee, apparently there have always been sufficient Swedish candidates for these committees. It is also permissible to appoint experts to take part in the deliberations and decisions of the Nobel committees.

Paragraph 7 stipulates that 'No person may be considered eligible for an award unless nominated in writing by a person competent to make such a nomination.' This sounds innocent enough; after all, why should the Nobel committee consider someone who had not been nominated? An example, however, illuminates the importance of this stipulation.[23] The American Moses Gomberg was nominated for the Nobel Prize in Chemistry in 1921 for the discovery in 1901 of the first free radical. One of the committee members made a report concluding that the German W. Schlenk had made a decisive contribution to the understanding and ascertaining of Gomberg's discovery. Thus Gomberg and Schlenk should share a prize for the free radicals. This, however, was not possible because Schlenk was not nominated in 1921 although he had been nominated in 1918 and 1920. Gomberg was first nominated in 1915; however, the nominating letter did not arrive in Stockholm in time. Nominations are valid if they arrive before 1 February. L. Chugaev sent his nomination from St. Petersburg, dated 12 January 1915. At that time the Russian calendar was still 13 days behind, and the czar's secret police further delayed the mail. For 1916, Chugaev was not invited to nominate and his 1915 nomination was no longer in effect. Eventually, Gomberg was nominated eight times[24] between 1921 and 1940 by a total of 16 persons, but he never received the Nobel Prize (and neither did Schlenk).

No appeal may be made against the decision of a prize-awarding body with regard to the award of a prize, as is stated in §10. The finality of the Nobel decisions makes it especially important that they be well researched and justified. Tremendous care, labor, and expense go into the decisions. It is typical for the child of a Nobel committee member to remember virtually fatherless summers when her father was closeted in his office surrounded by papers, journals, and books, working on his report for the September committee meeting. In spite of the numerous lectures and seminars, Swedish scientific life may be incompletely informed, may be prone to undue influence by great foreign scientific authority, and may be subjected to manipulation by skillful lobbying.

Biases

Lobbying can also serve a useful purpose. It can bring new information to light and augment existing information for the Nobel committees. Some foreign scientific attachés in Stockholm are on the lookout to forge useful contacts and further their national candidates' chances for Nobel attention. Lobbying for different candidates may be combined if an integrated effort promises a higher probability of success. Such actions are supposed[25] to have worked for the Nobel Prize shared by the Japanese Kenichi Fukui and the American Roald Hoffmann (in 1981) for their theoretical studies of chemical reactions. Similarly, lobbying is supposed[25] to have happened in the case of the 1987 chemistry prize awarded jointly to the Americans Donald Cram and Charles Pedersen and the French Jean-Marie Lehn for their development of supramolecular chemistry. Joining forces does not necessarily mean joint nominations. Simultaneous but separate nominations of two candidates in the same field may strengthen each other by focusing attention to their field. Such co-ordination of nominations is not unethical although it may be against the rules. The letters of invitation to nominate contain a stern warning not to discuss the nomination with others.

High-level lobbying may include dissertation-length treatises like any other scientific documentation. The Swedish judges welcome such efforts because it makes life easier for them. On the other hand, it does not diminish their responsibility because no matter how thorough such documentation is, it may still be biased or negligent toward other candidates, inaccurate in the question of priority, and so on. In any case, lobbying is serious business, best done on a continuous basis. A Swedish scientist compared it to chromatography, in which the substance is put into the column in the right form, and then you have to keep adding solvent to it. Sometimes your substance sticks in the middle of the column and won't come out, but you continue pouring solvent on it and eventually it may come out. Swedish scientists may meet undue politeness and reverence when visiting foreign laboratories, but they are conscious of this and take it in their strides. Of a Swedish and non-Swedish scientist of equal scientific standing the Swedish will have an advantage in getting an invitation, an honorary degree, and suchlike if the people involved have any concern about the Nobel Prize. Gösta Ekspong,[26] a distinguished professor of Stockholm University and long-time former member of the physics Nobel committee, has experienced how eager people may be to show kindness towards Nobel committee members, but he was equally eager not to be influenced by their kindness.

When Lars Ernster[2] (1920–98), later a distinguished biochemistry professor of Stockholm University and long-time Nobel committee member, was preparing for his Master's examination in Stockholm, his examiner gave him a special assignment of preparing an overview of David Keilin's contributions to the elucidation of the mechanism of cell respiration. Keilin was a pioneer biochemist in Cambridge. Ernster's professor wrote a letter to a leading British biochemist for information and within days received a nine-page evaluation of Keilin's scientific achievements.[27] The request from Sweden for material to augment a Master's examination may have been read as something more, and could have been considered as a request in disguise for Nobel evaluation.[28]

The Nobel Archives open their files for research only after 50 years have elapsed. This was introduced in 1974 in the *Statutes*. Alas, the information contained in the archives is rather limited. There are the letters of nomination, the reports of Nobel committee members and outside experts, and the general reports reviewing all nominations. Missing, however, are all the deliberations because, under rigorous rules, they cannot be recorded.

The records do show, for example, that the unwritten rule of the 20 years' grace period for the recency stipulation was observed in handling Gomberg's case. Lately it has not been followed. The fact that Chugaev's late nomination could not be considered the following year seems rather strict. In later years this has been relaxed; when a nomination reaches the Nobel committee after 31 January it is automatically added to the nominations for the next year. At the end of 1954, documentation on Nikolai Semenov's (C56) scientific activities was sent from Moscow to Stockholm.[29] Apparently the Russian postal services had not improved much during the preceding four decades because the documents did not arrive in time for consideration for the 1955 Nobel Prize. They were taken into account in the selection process for the 1956 prize.

It is hardly possible even to imagine a prize-selection operation without any bias. The question is how strong may such a bias be, how can the juries safeguard their institution against it, and how objective can their judgements be? Up until the Second World War, Swedish science had much stronger ties with German science than with American and British science. Swedish scientists were not alone in having great respect for German science in earlier times. Around the turn of the twentieth century it was customary for American and British chemists to get their doctorates in Germany.[30] Up until the 1930s, physicists were better off if they could read German because the most important papers appeared in that language. Such giants of American science as the chemist Linus Pauling (C54, Peace 1962) and the physicist Robert Oppenheimer spent postdoctoral years in Germany learning from the best people in their fields. When reading the special report of O. Widman to the Nobel Committee for Chemistry, evaluating Gomberg's discovery, one cannot miss the adulatory tone in Widman's description of Schlenk's contribution, a tone which is missing from his account of Gomberg's work.[23] Widman writes, 'In this year W. Schlenk started to publish his *masterly* studies, emanating from the *famous* Munich laboratory' (my italics). Widman himself spent a period of his travels in his late twenties, around 1880, in Adolf von Baeyer's (C05) organic chemistry laboratory in Munich.

Human bias is not easy to avoid. Had Widman been more sympathetic toward Gomberg, it might have been easy to convince his colleagues that Gomberg deserved the Nobel Prize. Instead, Widman made much of Gomberg's hesitation in recognizing his own discovery as proof of the first free radical. This hesitation was expressed in just two sentences of Gomberg's 400-page writings on the topic, and his doubt lasted for only a few months. Nonetheless, Widman invoked it every time the question of the prize for Gomberg came up.[23]

The small Nobel committees, consisting of five members each plus the secretary, can scarcely cover all the relevant fields and have to rely heavily on the expertise of their colleagues and outside specialists. Lars Ernster said:

> What sitting on the Nobel Committee gave me most of all was
> that I tried to understand and developed a respect for other fields,
> different from mine. It was also obvious that we should not try to
> become experts in other fields, to the extent that we could make
> judgments. It would be both pretentious and futile. What I tried
> to do was to understand, appreciate, and respect the opinion of
> my colleagues.'[2]

Nobel committee members have observed that once someone has been engaged in preparing reports on certain fields and candidates, he often becomes an advocate of those fields and persons. This is not surprising since the committee members are assigned to the fields that are the closest to their own interests. There is then an additional incentive. The committee member whose 'charge' is finally selected for the prize is often the one to make the presentation at the award ceremonies on 10 December, in the presence of the king, the TV cameras, and everybody else.

Nikolai Semenov's Nobel Prize shows how an individual, even if not a committee member, may achieve a lot in Nobel matters.[31] Lars Gunnar Sillén was Professor of Inorganic Chemistry at the Royal Swedish Institute of Technology in Stockholm from 1951, at 35, until his early death in 1970. He worked to get documentation on Semenov's scientific activities for the Nobel Committee from 1952 until 1954, when he finally succeeded. He was devoted to improving Swedish–Soviet relations and he knew the Russian language. Much more conspicuous roles played by individuals in Nobel matters are also known from the early history of the physics and chemistry prizes. Gösta Mittag-Leffler, the mathematician, had a strong though not always successful influence over the physics prizes, and Svante Arrhenius (C03) had a mostly successful impact on both physics and chemistry prizes.[32]

Sometimes the Nobel committees are accused of being influenced by politics. The early German orientation of Swedish science has already been mentioned. In an ironic twist, after Hitler in 1937 had banned German nationals from receiving the Nobel Prize, three German nationals received awards in 1939 (see next chapter). Although Germany tried to discourage the Swedish prize awarders from giving the Nobel Prize to German scientists, a member of the chemistry committee explained that no government action could influence their choices. This member was Hans von Euler-Chelpin (C29), professor at Stockholm University and German by birth. He was known as a Nazi sympathizer and he concluded his letter with 'Heil Hitler, Mit deutschem Gruss.'[33] Possible political consideration of a different kind was brought up by some in connection with the prize for atmospheric chemistry in 1995. Of course, the Swedes have denied that politics might have played any role in it. Realistically, though, concerns about the environment have increased recently and become a major issue. If science can contribute to solving problems of the environment, giving a Nobel Prize for such a contribution corresponds to Nobel's original intentions as much as anything else.

Political attacks were part of the atmospheric chemists' lives. Their findings were unwelcome to some of the major chemical companies. Rowland (C95) notes that

> In the 1974 time frame, two thirds of the CFC [chlorofluorocarbons] production was in the form of propellants for aerosol sprays in the United States. About half of the global CFC use was then in the United States. The aerosol spray propellants were mostly chlorofluorocarbons. Some used hydrocarbons, but probably 80% of the aerosol industry were in the front line immediately, and they came out swinging. In the American Western movies, the good cowboys always wear white hats, and the bad guys always wear black hats so the little kids know whom to cheer for. We had black hats, as black as you could get. Every month Molina and I would read in their publication, *Aerosol Age*, to find out what else they were saying about us. One that illustrates to me how far they would go was when they had an interview with a person who speculated that we were agents of the KGB,[34] intent on disorganizing American industry.[35]

To the extent that the issue of the environment can be considered a political question, Nobel decisions may be influenced by politics. However, taking politics in a less liberal sense, political bias in Nobel matters is 'totally unthinkable'. This is the vehement denial by George Klein, emeritus professor of tumor biology at the Karolinska Institute. He is a long-time former member and one-time chairman of the Karolinska Nobel Assembly. He maintains[9] that the scientists involved are not affected by politics, that they could not care less about ideology. He illustrates how unthinkable it is with a story. The chief rabbi and the Catholic cardinal are having dinner together and the rabbi refuses to eat non-kosher food. The cardinal tries to convince him to try and the rabbi gives in, 'All right, I'll eat it on another occasion.' The cardinal asks him, 'When?' The rabbi answers, 'At your wedding.' Klein's reaction is the more remarkable because he observes strong political influence and the strong impact of personalities in other aspects of Swedish life. A case in point is when Swedish journalists and diplomats were misled by the infamous Pol Pot regime in Cambodia about its intentions and practices of genocide. This was compounded by instructions from the Swedish government, which found it advantageous to cast the Pol Pot regime in a favorable light and thus strengthen anti-American sentiments at the time of the Vietnam War.[9] Accusations of political influence in Nobel matters may come from diverse corners and causes. When Albert Einstein (P21) received the Nobel Prize, Philipp Lenard (P05) protested that the Swedish scientists 'had not been able to bring to bear a sufficiently clear Germanic spirit to avoid perpetrating such a fraud.'[36]

Nominations

Broad international participation in the nomination process can do much to compensate for dangers that may emanate from the inbred character of the small and exclusively Swedish Nobel committees. Nowadays, thousands of letters soliciting nominations are mailed out each fall, and hundreds of responses come in before the 1 February deadline. It is yet another source of the high prestige of the Nobel Prize that the international scientific community is being involved in the procedure. If the Nobel awards went to people who did not enjoy the respect of the international community, the esteem in which the prize is held would soon wane. In fact, the opposite is true.

Those who may make nominations are outlined below. There is permanent entitlement in four categories, and there are two categories of ad hoc nominators.

Nominations for the physics and chemistry prizes can be submitted by

(1) Swedish and foreign members of the Royal Swedish Academy of Sciences;

(2) members of the Nobel Committees for Physics and Chemistry;

(3) physics and chemistry Nobel laureates;

(4) permanent and assistant professors of physics and chemistry in Sweden, Denmark, Finland, Iceland, and Norway;

(5) professors of physics and chemistry in selected schools in other countries (the selection changes from year to year);

(6) other scientists who receive individual invitations to submit nominations.

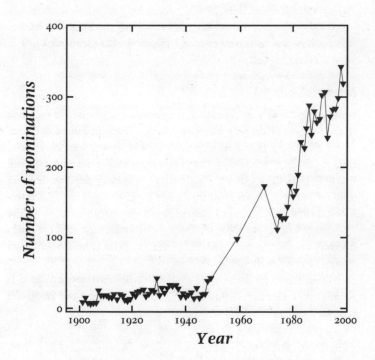

The dramatic growth in nominations for the chemistry Nobel Prize
(courtesy of Lennart Eberson)

Nominations for the physiology or medicine prize can be submitted by

(1) members of the Karolinska Assembly, comprising 50 elected members from among the professors of medical subjects at the Karolinska Institute;

(2) Swedish and foreign members of the medical class of the Royal Swedish Academy of Sciences;

(3) Nobel laureates in physiology or medicine;

(4) professors of medicine in Sweden, Denmark, Finland, Iceland, and Norway;

(5) professors of medicine in selected schools in other countries (the selection changes from year to year);

(6) other scientists who receive individual invitations to submit nominations.

At the end of January, when all the nominations are together for that year, the Nobel committees review them. Their members have the right to nominate; thus they submit names, usually in the form of one-sentence recommendations, of people they feel should have been there but for which no nomination had arrived. The Nobel Archives contain many such last minute nominations, some more conspicuous than others. Thus, for example, in 1943, a committee member[37] nominated Otto Hahn and another committee member[38] nominated Artturi Virtanen, both in one-sentence letters.[39] In 1944 and in 1945, again, one-sentence nominations are entered on 31 January for Hahn.[40] Virtanen got several nominations in 1945.[41] The Nobel Prize in Chemistry for 1944 was awarded to Otto Hahn for the discovery of nuclear fission, and for 1945 to Artturi Virtanen for his fodder preservation method. Both were awarded in 1945. Hahn's prize was controversial, because it came so soon after the Second World War and the first atomic bombs, and because it was not shared with Lise Meitner (see p. 232). As for Virtanen's prize, it has been alleged[42] that it was politically motivated by pro-German and anti-Soviet sentiments. The presenter at the award ceremony began his address to Virtanen (a Finn) with the following words: 'I do not think I am wrong to say that it is in your ardent patriotism that we must look for the most powerful force inspiring your great scientific achievement.'[43]

On other occasions a forceful nomination from a great authority seems to carry a decisive influence. Albert Einstein's laconic telegram of 19 January 1945[44] reads: 'nominate wolfgang pauli for physics-prize stop his contribution to modern quantumtheory consisting in so-called pauli or exclusion principle became fundamental part of modern quantumphysics being independent from other basic axioms of that theory stop albert einstein.'[45] Pauli received the Nobel Prize in Physics for 1945, and the citation followed Einstein's formulation: 'for the discovery of the Exclusion Principle, also called the Pauli Principle'.[46]

A curiosity of the 1946 nominations was that three committee members[47] submitted nominations in the last days of January 1946, for W. M. Stanley, J. B. Sumner, and J. H. Northrop, and these indeed shared the 1946 chemistry prize.

Secrecy is an important consideration in Nobel matters. There is a stern warning in the letter of invitation for submitting nominations to act in secret: 'You should treat your invitation and nomination as highly confidential and should not discuss them.' The Nobel committee is interested in the nominator's personal views rather than in the views of collective bodies. This secrecy helps to protect the nominators from undue pressure by potential Nobel laureates and their friends.

Although the circle of nominators is rigorously defined, it does not mean that others cannot in some way influence the Nobel committees. Whereas nobody outside this circle can submit official nominations, anybody can write letters to members of the Nobel committees and they are interested in any reasonable suggestion. There is always the danger of overlooking someone important or a poorly publicized discovery. The committee members serve as a filter and depository of suggestions.

After the closing date for nominations, 31 January, the next step is the hard work in the committees, involving experts. Rigorous secrecy surrounds their activities, protecting the members from undue diversions. The Science Academy and the Karolinska have somewhat different mechanisms once the committees have finalized their recommendations. At the Academy the recommendations go to the respective classes, the physics class and the chemistry class. The final vote is taken at the general meeting, in which all members participate. At the Karolinska the recommendations go to the Nobel Assembly, consisting of 50 professors of the Karolinska, which then makes the decision for the physiology or medicine prize. On the day of the final vote, there is a one-hour lecture in each subject, presenting the state of affairs of that subject and discussing in detail the merits and drawbacks of all noteworthy recommendations.

Toward the end of the process, as the time for the announcements approaches, secrecy increases expectations and tension. The Nobel announcements are made immediately following the votes at the Science Academy and in the Nobel Assembly at the Karolinska. The members of the Academy are discouraged from leaving the building before the press conference lest anybody leak the news. Although nobody could realistically think that the selection process has not ended some time before this, officially the decision is made by these final votes and, theoretically, the original recommendations could be overturned. Once the decision is made the new laureates are notified, then the general public. However, there is no strict rule about this; rather, it is a courtesy, and if the laureate cannot be located, the public announcement follows anyway. At that moment, all the information about the new Nobel Prize becomes available on the Internet. The Internet announcements[48] have all the works, references, brief biographies, description of the scientific significance of the discoveries, photographs of the laureates, and elaborate illustrations. They imply a long and laborious preparation, and participation in it by at least a few people.

Since the Internet has become involved, activities in the virtual Nobel world have mushroomed. Information not only about the new prizes and the laureates but also about all previous prizes has become available. One can read the speeches and lectures of the laureates of any year and any field. The services that are provided expand along with the technological advances, and the most recent award ceremonies can also be viewed in motion. The Nobel e-Museum (NeM) is fast becoming an encyclopedia, a TV station, and an interactive science workshop all rolled into one. On the day when the physics and chemistry prizes are announced, there are well over a million hits at the NeM site.[49]

War times

During both world wars there were some years when the prizes were not awarded. After 1939, Nobel Prize announcements were not made again until 1944, but wartime conditions hindered travel. Isidor Rabi's Nobel Prize in Physics for 1944 was announced in November of that year and he acknowledged it immediately in a telegram. However, he could not go to Stockholm, and in a letter of 28 February 1945 he apologized to the Swedish Academy of Sciences for not being able to come and give his public lecture. In the event it was decided not to wait, and the Academy transferred Rabi's prize money, diploma, and medal to the Swedish Embassy in the United States. The award ceremony took place at Columbia University in New York on 11 April 1945.[50] The president of Columbia, the 83-year-old Nicholas M. Butler, presented the diploma and medal to Rabi, who had received the money, USD 29 000, earlier. Butler himself was a Nobel laureate in peace for 1931 (for promoting the Briand–Kellog Pact). This occasion resembled an event five years before, in which Ernest Lawrence (P39) of Berkeley had refrained from going to Stockholm to the awarding ceremony in December 1939 for fear of the German U-boats. The Swedish consul in San Francisco brought the citation and the medal to the ceremony, held in Berkeley on 29 February 1940, which was presided over by the university president Robert Sproul.[51]

Otto Stern, whose Nobel Prize in Physics for 1943 was announced together with the 1944 prizes, and who had emigrated to the United States from Germany, could not make the trip either. The chemistry prize for 1943 was also announced in 1944 and the winner, George de Hevesy, conveniently, was living in Stockholm at the time.

In his letter of 28 February 1945 to the Swedish Academy, Rabi stated: 'It was a very remarkable and courageous act to resume the awards of the Nobel Prize in the midst of this bitter war. It was widely regarded on this side as still another example of the devotion of the Swedish people to the profound humanitarian ideals as exemplified by the life of Alfred Nobel.'[52] Of course, a fresh Nobel laureate's evaluation of Sweden's role in the Second World War would be different from, say, that of a Norwegian resistance fighter who watched with anger the trainloads of Swedish steel products destined for Nazi Germany throughout the war years. Leafing through the folders of the war years in the Nobel Archives, I had mixed feelings. Although the prizes were reserved during the first years of the war, otherwise everything seemed business as usual at the Science Academy and the Nobel Foundation during all those fateful years.

If the Nobel Foundation was not fully aware of the ravaging war going on in the rest of the world, there was plenty of indication of it in the responses to the invitations for nominations for the Nobel Prize. The dean of the Massachusetts Institute of Technology warned: 'Most of the important investigations are of confidential nature, and it is my belief that a satisfactory selection of a candidate could not be made until the war is over.'[53] Archibald Hill (M22) wrote, 'I feel that it would be almost a mockery to award the Nobel Prizes and to hold the Nobel celebrations.'[54] Professor Florence's (Lyon, France) wife apologized:

> I have the deep regret to tell you that my husband Professor
> FLORENCE is at present unable to avail himself of the honour
> you do him by asking him to nominate a candidate for the Nobel
> Prize for Chemistry for 1945. After being kept for some time in
> prison in Lyon, my husband has been deported to Germany. I have
> not heard from or about him since July; I only know he has been
> sent, with other French University Professors, to a concentration
> camp near Hamburg.[55]

Other prizes

The disproportionately high prestige of the Nobel Prize puts a grave burden on its awarders. It would be better to have a gradual gradation of prizes so that the Nobel Prize did not carry with it such a watershed effect. There are many other international science prizes,[56] and some go with large sums of money, too, yet none has come even close in prestige. We mention only a few here that have some relevance to the Nobel Prize.

The Crafoord Prize is the most important science prize in Sweden after the Nobel Prize, with half its monetary value, and covers fields that the Nobel Prize does not. It is presented by the Swedish king at a ceremony each fall, without the festive circumstances of the Nobel Prize. The Crafoord Fund was established in 1980 by a donation to the Royal Swedish Academy of Sciences from Anna-Greta and Holger Crafoord. The prize is international, awarded annually, and can be divided among up to three recipients. The prizes are awarded on a rotational basis in the following fields: mathematics; geosciences; biosciences; and astronomy. In addition, there may be a prize in rheumatoid arthritis but only when a special committee shows that scientific progress in this field has been such that an award is justified.[57] The Crafoord Prize in the biosciences has considerable overlap with the Nobel Prize in physiology or medicine.

The president of the State of Israel presides over the annual award of the Wolf Prizes in chemistry, mathematics, medicine, physics, and the arts.[58] The ceremony is held in the Chagall Hall of the Knesset (parliament) building in Jerusalem before hundreds of guests. The award of the Wolf Prizes was begun in 1978 by Ricardo and Francisca Wolf. Candidates are nominated by worldwide invitation to heads of institutions or departments in the respective areas, by previous recipients of a Wolf Prize, and by scientists or artists invited to submit nominations by the Wolf Foundation. The decision in each area is made by a committee of three, including one member from North America, one from Europe, and one from Israel. These committees are chosen by the council of the foundation and are appointed for a single year, with one member being carried over to the next year. On several occasions the winner of the Wolf Prize has then been awarded the Nobel Prize.

The Lasker Medical Research Awards of the Albert and Mary Lasker Foundation in the United States seem also to have relevance to the Nobel Prizes.[59] In spite of their relatively low monetary value, they carry high prestige. Albert Lasker was an advertising executive, and some ascribe the extraordinary success of his prize to his exceptional abilities in his profession. Three awards are made every year: the Basic Research Award, the Clinical Research Award, and the Special Achievement Award. The Nobel Prize in physiology or medicine is primarily aimed at basic research and only seldom at clinical research, and there is a considerable overlap in winners of the Albert Lasker Basic Medical Research Awards and Nobel Prizes. As with the Wolf Prize, a Nobel Prize often follows the Lasker Award. One might wonder whether the Lasker jury tries to anticipate the Nobel decisions or whether the Nobel committee relies on the Lasker decisions. Some of the previous Lasker awardees would now get a Nobel Prize in chemistry rather than in physiology or medicine, in keeping with the recent shift in the chemistry Nobel Prizes towards biochemistry. Only on one occasion (in 1960) was a Lasker Award followed by a Nobel Prize in physics, to Ernst Ruska in 1986, for the electron microscope. A survey of the period between 1964 and 1996 shows that there were only five years[60] when a Lasker Award did not go to at least one future Nobel laureate. On the other hand, only seldom has a Nobel laureate later been given a Lasker Award. One such was Albert Szent-Györgyi (M37), who received the award in 1954 for his muscle research, work that was unrelated to the research for which the Nobel Prize had been given to him. Then, Frederick Sanger (C58, C80) shared a Lasker with Walter Gilbert (C80) in 1979. Sanger had his first Nobel Prize, but his Lasker Award preceded the second. The mechanism of the selection of the Lasker awardees is different from the Nobel system. The Lasker has a large committee, with international representation, of highly visible scientists who work for a short period of time. The Lasker Prize is more American than international, yet prestigious if only by the sheer volume of American science. Although it is thought to be prone to influence by national interests, a counter-argument is that such national interests are less expressed in the United States than in, for example, European countries. In the United States regional interests or interests of particular schools may be important.

The institution of the Nobel Prize has been studied and emulated, which is a sure sign of recognition. An essentially reverse Nobel Prize has also been created. The magazine *Annals of Improbable Research* bestows the Ig Nobel Prize to individuals whose achievements 'cannot or should not be reproduced'. The Ig Nobel Prizes are handed out annually by genuine Nobel laureates in Cambridge, Massachusetts.[61] The Ig Nobel Prize is organized by academics *for* academics, rather than *against* pseudo-science or anti-scientists.

The first chemistry Ig Nobel Prize was awarded to Jacques Benveniste for his discovery that water has a memory and can thus carry molecular information. In 1998, he received a second Ig Nobel Prize for discovering that this molecular information can be transmitted over telephone lines, allowing homeopathic remedies to be delivered by e-mail.[62] The 1996 physics Ig Nobel went to Robert Matthews of Aston University, England, for his studies of Murphy's Law, and especially for demonstrating that toast more often falls on the buttered side. The 1998 medicine Ig Nobel was awarded to several people in charge of tobacco companies 'for their unshakable discovery, as testified to the U.S. Congress, that nicotine is not addictive.'[63]

The uniqueness of the Ig Nobel Prize is that often the awardees are outstanding scientists who have made their names with other discoveries. Benveniste, for example, is credited with being one of the discoverers of the so-called platelet-activating factor. This is a potent biological phospholipid, which causes the release of serotonin from platelets and is effective at extremely low concentrations. It is also a major signaling molecule in inflammatory and allergic conditions. Its discovery might be worth a real Nobel Prize.[64]

2

The Nobel Prize and national politics

C. V. Raman won the Nobel Prize in Physics in 1930 for the discovery of the Raman effect. He once said that although scientists 'are claimed as nationals by one or another of many different countries, yet in the truest sense, they belong to the whole world.'[1] Raman studied physics in India and got his degrees from the University of Madras. He did not follow the usual route of Indian scholars and go to Britain to get a PhD degree. He worked in a government office and in 1917 accepted a physics professorship at the University of Calcutta. He and his associate, K. S. Krishnan, observed a phenomenon of light scattering, first with their naked eyes, later with some optical instruments. They communicated the discovery in 1928. Legend has it that as soon as Raman had submitted the manuscript on the new effect for publication, he booked his passage to Stockholm from Calcutta for the following December. This is believable to the extent that Raman was known to be self-confident and that his Nobel recognition followed his discovery with unusual speed.

When the Russian Boris Vainshtein evaluated the British crystallographer Dorothy Hodgkin's (C64) life and works, he noted that she 'has done much for the glory of her homeland'.[2] This reflected more Vainshtein's way of thinking than Hodgkin's aspirations.

Raman was Indian and India could rightly claim him. Many of the Nobel laureates, however, migrated from their native lands, often changing countries more than once. This migration comes naturally to scientists, even under the most peaceful and ordinary conditions. It can be likened to the obligatory wanderings of European artisans in the Middle Ages, which were aimed at picking up experience in distant lands. In the 1930s a mass migration of another sort took place. From Germany and other countries in central Europe, scientists, mostly Jewish, sought refuge in Great Britain and the United States. Since the Second World War, too, there has been a steady migration of scientists responding to the pull of better research and economic opportunities afforded primarily in the United States and Western Europe.

Citizens of the world

It is a complicated matter to assign Nobel laureates to specific countries. Their country of birth, education, graduate studies, the country where they performed their prize-winning research, or the country where they lived when they received the Nobel Prize may all be criteria. When David H. Hubel (M81) was born in Windsor, Ontario, he acquired United States citizenship through his parents and Canadian citizenship by birth. In his Nobel autobiography he wondered whether each country would get half a credit from his becoming a Nobel laureate or both countries would get full credit.[3] Aaron Klug (C82)[4] was born in Lithuania when it was an independent country and left it with his parents when he was two years of age. The Lithuanians never claimed him as their Nobel laureate, but the Soviet Academy of Sciences once did when Lithuania was part of the Soviet Union. There is no unambiguous definition for the nationality of Nobel laureates. Often the country that had persecuted them and kicked them out later claims the Nobel laureate as her own. There are ways of juggling the data but the lesson is hard to avoid 'that science is far more international than flag-wavers care to acknowledge'.[5]

Some trends are interesting to observe. There are conspicuously strong countries in exporting and others in importing future Nobel laureates, based on the comparison of the birthplaces of the laureates and their domicile at the time of their awards.[6] The emerging pattern is that Germany, Austria, Canada, Hungary, Italy, and Poland, in decreasing order, have exported the largest number of Nobel laureates. The United States and Great Britain were the two largest importers, in a 3:1 ratio. Further patterns emerge, such as the pre-eminence of American science during the second half of the twentieth century. It is interesting that Cambridge University can claim more Nobel laureates than any countries, apart from the United States, Great Britain, and Germany. Many American universities also have enviable records though there, again, assigning a Nobel laureate to a particular school may be complicated. Joshua Lederberg (M58) found himself in an awkward situation when his Nobel

Prize came while he was in transition between the University of Wisconsin and Stanford University.

When Herbert Brown of the United States and Georg Wittig of Germany shared the Nobel Prize in Chemistry in 1979, the *US News & World Report* saw in the splitting of the prize a plot to curtail the growing American dominance in the Nobel Prizes. Reporting the 1979 Nobel Prizes in Physics and Chemistry, *The New York Times* headlined, 'Americans at Harvard and Purdue win—German and Pakistani cited.' The physics prize went to Sheldon Glashow and Steven Weinberg of Harvard, and to the Pakistani Abdus Salam of Trieste (Italy) and London.[7] When John A. Pople shared the Nobel Prize in Chemistry in 1998, the British press stressed that he was the third British scientist in as many years distinguished by this award. Pople had left Britain for the United States 34 years before. Although he had held a comfortable position in Cambridge, the United States provided a much greater opportunity for his work in computational chemistry, which was then an emerging field.[8]

Ernest Walton shared the Nobel Prize in Physics for 1951 with John Cockroft 'for their pioneer work on the transmutation of atomic nuclei by artificially accelerated atomic particles'. This work was done at the Cavendish Laboratory in Cambridge. Eventually Walton moved to Ireland, to Trinity College in Dublin. When he received the news about his Nobel Prize, he commented, 'The news of this signal honour has given great satisfaction not only to Trinity College, Dublin, but indeed to people all over Ireland.'[9]

The *Washington Post* of 11 October 1996 reported the physics and chemistry prizes with the following headline: '5 Yanks Win Nobel Prizes'. There was then a headline in a London daily reporting the buckminsterfullerene award in 1996 as 'Kroto scored one for Britain'.[10] This exemplified a new level of national awareness of the Nobel Prize in Great Britain. According to Aaron Klug,[4] president of the Royal Society, 1995–2000, this had happened only recently because in earlier days the British people used to pay little attention to the Nobel Prize. It contrasted with the Continent, where these things were regarded as terribly important. In the imperial days the British just took it for granted that their science was of high quality, but they afford it greater respect now that Britain has become a moderately sized European nation. The changing emphasis was there when Nevill Mott (P77) complained in the press about budgetary problems for science in the United Kingdom, and his article appeared under the title, 'Rocking the Cradle of Britain's Nobel Prize Babies.'[11]

Spain does not fare well among the European nations for the number of Nobel laureates. Santiago Ramón y Cajal (M06) shared an early prize with the Italian Camillo Golgi for 'their work on the structure of the nervous system'. Severo Ochoa (M59) was born in Spain, and received his education and degrees there, and his interest was greatly stimulated by the works of Ramón y Cajal. However, in 1936 he left Spain for good, and eventually settled in the United States where he carried out much of his scientific work.[12] According to José Elguero, former head of the Spanish Research Council, Spanish science was left behind in the nineteenth century as compared with France or Great Britain. The reasons go farther back in history:

> Spain used to be the fighter for the Catholic Church all over the place. The Spanish King used to be the main warrior. The Jews were expelled from Spain in 1492, and this was an important factor because part of the culture was in the Jewish community. I don't know why, but it's certainly true that we have nobody to compare with Galileo, Newton, Gauss, or Leibniz.[13]

Elguero is apprehensive of the present European development that seems to favor a division, according to which science and industry concentrate in some countries, such as Germany, while people go to Spain for holidays or to retire. Currently Spanish science is reviving, but if support gets depleted, the results may fast disappear. Because it lacks a solid tradition, the scientific establishment is not very strong.

Japanese examples

The Japanese Leo Esaki (P73) wrote, in his 1998 obituary for Kenichi Fukui, then Japan's only Nobel laureate in chemistry (1981), that Japanese scientists should earn more Nobel Prizes so that Fukui could rest in peace in his grave.[14] Soon enough, in 2000, Japan had another chemistry Nobel laureate. The Japanese take the Nobel Prize seriously. They consider Sweden as a supreme judge of world science. The story circulates that when, a few years ago, the budget for scientific research was doubled in Japan, a Japanese delegation was dispatched to Sweden with this information and to convey their expectation of more Nobel Prizes. Given the country's population and economic might, there have been relatively few Japanese Nobel laureates, a total of six in the sciences, through 2000. They were Hideki Yukawa (P49), Sin-itiro Tomonaga (P65), Reona (Leo) Esaki (P73), Kenichi Fukui (C81), Susumu Tonegawa (M87), and Hideki Shirakawa (C2000).* Three of the six had an American affiliation when the award materialized. Many in Japan and elsewhere tend to see in the number of Nobel Prizes an expression of the degree to which Japan has become integrated into the world system of science. Hideki Yukawa wrote to A. Westgren, on receiving his telegram about the 1949 physics Nobel Prize:

> It is not only the highest conceivable honor for me as a scientist,
> but also the best stimulus to encourage all Japanese people for the
> establishment of a peaceful and civilized country. I am sure that all
> Japanese are deeply impressed by this unexpected happy news and
> are very grateful to your country.[15]

The small number of prizes for Japanese scientists may also be an indication of the timidity of Western science towards Japan. A recent scholarly treatise by J. R. Bartholomew[16] examines three cases in which Japanese scientists could have won Nobel Prizes.

The German Emil von Behring (M01) won the first Nobel Prize in physiology or medicine for his work on serum therapy and its application against diphtheria. He did his research in Robert Koch's (M05) laboratory in Berlin, where Shibasaburo Kitasato (1852–1931), from Japan, also worked for six years. Kitasato had come to the idea of serum therapy before Behring applied it to diphtheria. All nominations, though, except one, proposed Behring for the prize. The exception suggested a prize shared between the two, and it came from Budapest. According to Bartholomew, the nominator's 'Hungarian ethnicity gave him a greater affinity with things relating to Asia than most European medical and scientific personalities in this period.'[17]

Katsusaburo Yamagiwa (1863–1930) could have shared the Danish Johannes Fibiger's (M27) prize. Fibiger ostensibly demonstrated that a parasitic worm could cause tumor in the stomach of certain rats. Later it was shown to be an artifact and that the tumor developed as a consequence

---

*In 2001, there was yet another Japanese laureate in chemistry,
 Ryoji Noyori of Nagoya University.

of vitamin A deficiency. On the other hand, Yamagiwa had conceived a simple and reliable methodology of generating tumors in rabbits by painting their ears with coal tar. Fibiger received many more nominations for the Nobel Prize than Yamagiwa, whose nominations came primarily from Japan. Not only was it unjust that Yamagiwa did not at least share this prize, which Fibiger received alone, but it has also become known as one of the most blatant blunders in Nobel history (see also p. 75).

Both Kitasato's and Yamagiwa's missing prizes are conspicuous because others working in the same areas did get the awards. Genichi Kato (1890–1979), on the other hand, worked in an area for which no Nobel Prize was given. His claim for the prize could have been the isolation of single muscle and nerve fibers, and the discovery of reflex inhibitory and excitatory nerves. In this case Japanese politics was more involved than in the two other examples. A high official in the Japanese foreign ministry,[18] was active in taking steps to facilitate Kato's candidacy, and Japan's minister to Sweden was also involved.[19] It is not known, though, whether this interference helped or hurt Kato's cause. In any case, there were roadblocks from Tokyo to Stockholm during the first half of the twentieth century. Six Nobel Prizes later there may still be reservations as regards Japanese science. John Maddox, the former long-time editor of *Nature*, thinks that the Japanese do not get their fair share of recognition for their work in science: 'I think people are bothered by the language and the culture, and most Westerners find Japan strange. The consequence is that Japanese scientists are not highly enough regarded.'[20] Whether it is popular but misguided perception or indeed has real foundation, the general impression is that where the scientist comes from counts when recognition is concerned. This is notwithstanding the fact that occasionally scientists from 'exotic' countries get the Nobel Prize. Usually, though, they have departed their country of origin years before receiving the award, and have worked for a long time in one of the 'in' countries, like the United States or Britain. This, for example, was the case with Abdus Salam (P79) of Pakistan and Ahmed Zewail (C99) of Egypt. I remember a party conversation in London about the possible reasons for the omission of Salvador Moncada from the 1998 Nobel Prize for nitric oxide research. When someone volunteered the information that the London-based FRS Moncada had come from Honduras, someone else exclaimed, 'Say no more.'

No prophet in his own country

Although most countries show great concern for having Nobel laureates, this is often not commensurate with their concern for scientists that may win the prize. Thus, for example, Fukui was not elected to the Japan Academy until he won the Nobel Prize, nor was he decorated with the highest Japanese award. Decorations may help scientists to attract attention and are one of the ways of lobbying for their international recognition. Of course, for this to happen the country needs to wield influence. A national award in Argentina, Hungary, or the Philippines would not mean a thing elsewhere, but this is not the case in a country like Great Britain or Germany.

Harold Kroto (C96) received a knighthood before the Nobel Prize, although most British Nobel laureates receive this distinction afterwards. Some decline it, such as Frederick Sanger (C58, C80), Francis Crick (M62), and Max Perutz (C62), while others have never been offered it, like Maurice Wilkins (M62), Brian Josephson (P73), and Antony Hewish (P74). Wilkins may have been left out because he is considered by many to have received the recognition that should have belonged to Rosalind Franklin (1920–58). She had died before the prize for the double helix was awarded. Josephson may have been side-stepped because of his engagement in studying paranormal phenomena and telepathy. In the case of Hewish, there was a controversy arising from the omission of Jocelyn Bell, a graduate student of Hewish who made the original observation of pulsars. Incidentally, while Sanger, Perutz, and Crick had declined the knighthood, they accepted the higher distinction of the Order of Merit, which is restricted to 24 individuals in the United Kingdom. Aaron Klug (C82) and James Black (M88) are also members of the Order of Merit.

The Nobel Prize often leads to other awards. The Order of Merit, of course, is very exceptional. As for lesser awards, one would think that it is almost a 'waste' to distinguish Nobel laureates with such awards and they could be better used by people who are not Nobel laureates. However, the award-giving bodies may be more concerned with the prestige of their award than with the distribution of distinctions among a larger number of scientists. A Nobel laureate's name certainly adds to the luster of any awardees' list. In 1996, for example, George Olah (C94) was awarded the F. A. Cotton Medal by the ACS Texas A&M Section, for the same achievements as in the Nobel citation.[21] Herbert Brown (C79) won the first Herbert C. Brown Medal, recently established by the American Chemical Society.

The Hungarian stamp issued for the centenary of Alfred Nobel's death listed 12 Hungarian Nobel laureates, considerably more than would be realistic. One of the dubious claims was 'Polányi János,'[22] that is, John Polanyi (C86). His parents were Hungarian but he was born and educated outside Hungary, never lived in Hungary, does not speak the language, and does not identify himself as Hungarian. The 1943 Nobel laureate in chemistry, George Hevesy, was listed justly because he was born and educated in Hungary. Later, though, he was practically kicked out of the country for his alleged political involvement in the short-lived Soviet government in Hungary in 1919. Hevesy, who was of Jewish extraction, then worked in Germany until he resigned under the Nazi regime and went to Denmark, where once again he had to flee along with the Danish Jews to Sweden. When the Germans occupied Denmark, Hevesy had dissolved the Nobel medals of Max von Laue (P14) and James Franck (P25), which had been placed in Niels Bohr's institute in Copenhagen for safekeeping, in aqua regia. The medals spent the war there, in bottles on a shelf. After the war the two laureates were presented with the recast medals.[23] This rivals the story of the great polymer chemist, Hermann Mark, who, when fleeing Nazi Austria carried his family fortune with him as coat hangers made of platinum. Somewhat later, in 1940, Howard Florey (M45) and his colleagues were gravely concerned about what to do with their strain of penicillin if the Germans were to invade Britain and reach Oxford. Determined to preserve the mold and not let the Germans discover it, they smeared the spores into the linings of their clothes where they would stay intact for years.[24]

Hevesy[25] was born as György Bischitz, in 1885, in Budapest as one of eight children of Lajos Bischitz and Jenny Schossberger. The turn of the twentieth century was an era of unprecedented prosperity in Hungary, following the formation of the dual Austro-Hungarian monarchy in 1867. In the enlightened atmosphere it was not uncommon for outstanding individuals in finance, industry, and trade to become members of the hereditary nobility, an honor conferred on Lajos Bischitz by Francis Joseph I, Hungarian king and Austrian emperor. The Schossbergers had been granted nobility even earlier, and were the first among non-converted

Jews to be so elevated. In Hungarian custom, nobility was signified by a pre-name before one's surname, indicating the geographical origin of the nobleman. Bischitz received the pre-name 'hevesy', after the region Heves. In 1906, the Bischitz family changed their foreign-sounding surname to Hevesy. It seems that Hevesy wanted to make sure that his name contained proper reference to his noble status and inserted 'von' in front of his name in his German publications and the French 'de' in those published elsewhere. He was not alone in such vanity. Other examples include the Hungarian–American mathematician, John von Neumann, and the Hungarian–American aerospace scientist, Theodore von Kármán.

In connection with Hevesy I used the euphemistic expression 'of Jewish extraction'. He did not consider himself Jewish yet he had to leave Germany at some point, and Denmark later during the Second World War, precisely because of his Jewish origins. There have been many Nobel laureates who had Jewish origins and who proudly considered themselves Jewish. Others may have been born Christian, or converted, or identified with no religion, and declared that they had nothing in common with Judaism. However, their lives, too, were often influenced by the fact that they had Jewish roots. Arthur Kornberg (M59), who has been sensitive to this issue remembers that this topic never came up in the conversations with his one-time mentors, the two Coris (M47). Carl and Gerty Cori were from Prague, and Gerty came from a Jewish family. The trauma of anti-Semitism forced them to leave Europe in 1920–21.[26] For some the distress caused by persecution that they or their parents had experienced made them break with their Jewish past upon arrival in America. This happened to many ordinary people. It is when a Nobel laureate considers his Jewish roots a taboo that it is especially conspicuous and sad.

The number of Jewish laureates is indeed very high. A recent book on the Nobel Prize tabulated an impressive list,[27] and in Russia a biographical dictionary of the Jewish Nobel laureates appeared recently.[28] The listings contain 116 names, of which there are 36 physics, 22 chemistry, and 39 physiology or medicine laureates, which means about one-fifth of all science laureates. The criteria for getting onto these listings are not well defined, and enhanced scrutiny would render such a compilation increasingly meaningless. The deep-rooted Jewish tradition that values education and knowledge, coupled with the persecution and discrimination from which, again, education could be a way out, are usually cited in explaining this phenomenon. Such reasoning implies that the United States and Israel would see a decline in the relative abundance of Jewish laureates a few generations after immigration. In this discussion I purposely leave out the possible genetic component as too speculative for the time being. George Klein (cf. p. 19) has been asked many times in Israel about the lack of Israeli Nobel laureates, in contrast with the many Jewish laureates elsewhere. Klein says: 'My answer is two words in Hebrew, "Atem babit", which means, "You are at home." A Jewish professor at Harvard feels that he has to run ten times as fast as a non-Jewish professor in order to convince himself that he is still standing in the same spot. That particular drive is not present in Israel.'[29]

This drive figures also in Leon Lederman's (P88) description of the ingredients of success in science, stressing that what is often considered to be a Jewish experience is not at all unique to Jews:

> You need certain elements such as strong family tradition,
> devotion to education, some healthy amount of insecurity to
> realize that you need to have hard work, and some drive for
> success. These are common elements of Asian families. By the
> third or fourth generation, they are no longer so different from all
> the other American families. It's lucky that there is always a new
> influx, and the next may come from Zaire or Brazil.[30]

Lederman's parents did not go beyond grade school. They both went from the Ukraine to the United States. It would be interesting to construct a map indicating all the places in Eastern Europe where Nobel laureates had originated. It might be one of the most densely populated areas as regards later Nobel laureates. Such a map would include regions that have changed hands repeatedly during the last centuries between Russia, the Ukraine, Poland, Czechoslovakia, Slovakia, Austria–Hungary, Hungary, Romania, and up north, Lithuania. The contrast between persecution in the old country and opportunity in the United States was staggering. The anti-Semitism in the United States during the first half of the twentieth century produced discrimination but not persecution, let alone extermination, and must have enhanced the drive and determination of the recent immigrants and their children. The road to Stockholm from Eastern Europe in most cases led through the United States and Great Britain.

Academy membership is probably the single most important national distinction a scientist may get because it comes from the scientist's peers. Every time someone who is not a member of his national academy gets a Nobel Prize it raises eyebrows, because this may indicate personal animosities or that the scientist is not well enough known at home. American Nobel laureates in this category include Elion (M88), Hauptman (C85), Nathans (M78), and Curl (96); all of them became members of the National Academy of Sciences (Washington, DC) swiftly, after they had won the Nobel Prize.

In 1965, three French scientists shared the Nobel Prize in Physiology or Medicine 'for their discoveries concerning genetic control of enzyme and virus synthesis'. None of the winners, François Jacob, André Lwoff, and Jacques Monod, was a member of the French Academy of Sciences. Age could not play a role in this as they were by then 45, 63, and 55 years old, respectively. However, Jacob[31] and his colleagues were rather hostile to the academy, which was an old system whose few members resisted change, and Jacob and the others did not want to announce their candidacy, which was a prerequisite. Finally, Jacob and Lwow were elected, on the initiative of others, but by then Monod had died, so that he was never a French academician. Then, in 1996, Jacob was elected to the most exclusive Académie Français, which has a total membership of 40, including only a few scientists; Jacob's task in the academy is to control the words used in biology and medicine.

Sometimes pressure mounts by the sheer accumulation of recognition. Barbara McClintock (M83) was not spoiled by distinctions during most of her career, but then, almost suddenly, she was showered with awards. These may have helped focus the attention of the Nobel committee on her achievements, or word may have gotten around about her anticipated Nobel Prize, triggering other awards. In the two-year period preceding her Nobel Prize she received more awards than during the rest of her career to that date. The Egyptian-born American Ahmed Zewail (C99) was showered with awards and other distinctions before his Nobel Prize, building up tremendous anticipation. His native Egypt issued a stamp set honoring him about one year before the Nobel Prize. Following the prize, another stamp was issued. Zewail noted 'I am particularly pleased as this honor comes from my country of birth and that I could be in the company of stamps honoring the pyramids, Tutenkhamen, and

Queen Nefertiti.'[32] The first set of Zewail stamps was rather unique because stamps are usually issued after rather than before the Nobel Prize is awarded to a living scientist.

Displeasing Hitler

Nazi Germany provided an extreme negative example of how far national politics may go with regard to the Nobel Prize. It banned its citizens from accepting Nobel Prizes in retaliation for the Norwegian Nobel Committee awarding the 1935 Nobel Peace Prize to the German pacifist and concentration camp inmate Carl von Ossietzky (1889–1938) in 1936.

In 1938, Robert Robinson (C47), a Briton, nominated Richard Kuhn, an Austrian who was working in Germany, for the Nobel Prize in Chemistry.[33] Robinson was aware of the German ban on the Nobel Prize and noted in his letter of nomination that although Kuhn was working in Germany, he was an Austrian national and the prohibition would not apply to him. His nominating letter was received in Stockholm too late to be considered in 1938. Robinson's nomination then figured among the nominations for the 1939 prize and the reserved 1938 Nobel Prize in Chemistry was awarded to Kuhn 'for his work on carotenoids and vitamins'. In the meantime, however, Germany had annexed Austria by the February Anschluss, and Kuhn had become a German citizen.

Half of the Nobel Prize in Chemistry for 1939 was awarded to Adolf Butenandt of Berlin 'for his work on sex hormones'. The other half went to Leopold Ružička of Zurich 'for his work on polymethylenes and higher terpenes'. The Nobel Prize in Physiology or Medicine for 1939 was awarded to Gerhard Domagk of Münster, Germany 'for the discovery of the antibacterial effects of protonsil', which was a sulfanilamid derivative.

The Nobel Archives at the Royal Swedish Academy of Sciences store some correspondence from Kuhn[34, 35] and Butenandt.[36, 37] Each sent a letter to the Academy in 1948 explaining what had happened nine years before. There is also a letter from each to von Euler of the Nobel Committee, sent in 1939, shortly after they had received the news of their respective Nobel Prizes. Apparently, in addition to the official telegram about the Nobel Prize, von Euler sent telegrams of congratulations to both Butenandt and Kuhn. In their responses, back in 1939, to von Euler's telegram, they thanked him and told him about their attempts to find out from the authorities what their official response might be. From their letters to the Academy in 1948 we can get some idea of what had then happened.

Butenandt and Kuhn were summoned to the Ministry of Education for instructions, 'Besprechung', and sat facing, across a table, a ministerial officer and another man that neither knew. Three typed letters, without heading, were placed on the table, letters that were to be sent to Stockholm, each one declining the prize and bearing a signature of one of the Nobel laureates. The third letter was presumably prepared for Gerhard Domagk (M39). When they tried to argue for correcting some factual errors, the unknown person exclaimed that 'every word has been approved by the Führer personally and was unalterable'. Both Kuhn and Butenandt, in their respective letters of 1948, explained to the Swedes that they were under pressure and could not have refused signing the letters without endangering, beside themselves, their institutes, their wives, and their five children (each). They both expressed the hope that their colleagues abroad were aware of the fact that they signed and sent their letters under duress. The Swedish consul in Frankfurt am Main handed the Nobel Prize diploma and medal to the laureates on 19 July 1949, almost ten years after the original announcement and four years after the end of the war.

Gerhard Domagk's fate was somewhat different at both ends.[38] Following the announcement of his Nobel Prize, he was arrested by the Gestapo and taken to prison for interrogation. After release he was also ordered to sign a letter declining the prize. In 1947, he traveled to Stockholm to receive the diploma and the medal from the Swedish king. The presentation speech[39] mentioned that during the eight years since the award announcement, Domagk's work had become even more significant.[40]

On the 50th anniversary of Kuhn's Nobel Prize, in 1988, Austria considered issuing a commemorative postage stamp, which would have coincided with a big international chromatography meeting in Vienna. But it was also the 50th anniversary of the Anschluss and a 1938 statement by Kuhn, in which he welcomed the annexation, surfaced. The plans for the stamp were shelved. It was issued four years later, for the 25th anniversary of his death.

Butenandt became a celebrated scientist in post-war Germany. He served as president of the Max Planck Gesellschaft, the most prestigious position in German scientific life. These days I address letters to my colleague in Munich at Butenandt Street. However, Benno Müller-Hill's book, *Murderous Science*,[41] documented some facts about Butenandt. When the infamous Josef Mengele was collecting samples of sera from prisoners in Auschwitz in his horrible experiments, he sent them to Berlin for analysis by a G. Hillmann, a guest co-worker at Butenandt's institute. When Müller-Hill confronted Butenandt about this, he denied any knowledge of it. Butenandt threatened to sue Müller-Hill, which he would have welcomed, but never did. Butenandt was on the committee of Hillmann's doctoral examination in 1947, and shortly afterwards Hillmann joined Butenandt's institute in Tübingen, where he had moved during the war. Butenandt also participated in the 1949 whitewash of Professor von Verschuer, who did research on 'materials' sent to him from Auschwitz by his former student Mengele.

There was a drop in German awardees following the Second World War, except for the controversial unshared 1944 chemistry prize to Otto Hahn. After the First World War, in contrast, there was no such drop. On the contrary, the awarders went out of their way to recognize German science, which did not decline so profoundly as a consequence of the First World War as between 1933 and 1945. Sweden had been a German-oriented society, much more so than, for example, Norway. Fritz Haber (C18) received his Nobel Prize in 1919 'for the synthesis of ammonia from its elements', work he had completed before the First World War. Haber's discovery and work greatly aided Germany's war efforts in the First World War. Besides, his name gained notoriety for his dedicated efforts to enable the German Army to use poisonous gases.[42] The chemistry committee and the Science Academy were much criticized for their choice, which was labeled as, at the very least, ill-timed.[43] Ironically, the strongly nationalist Haber died as a Jewish exile after Hitler's ascension to power.

Soviet dilemmas

The Soviet government also made it impossible for some of the Soviet Nobel laureates to travel to Stockholm or Oslo to receive their Nobel Prizes. The poet Boris Pasternak (Literature 1958) was one example and the physicist Andrei Sakharov (Peace 1975) another. The Soviet authorities let it be known that if Pasternak were to go to Stockholm to collect the prize, he might not be allowed back home. In Sakharov's case they prevented him outright from going to Oslo. His wife, Elena Bonner, went in his place.[44]

There is another interesting story of Soviet politics being mixed with Nobel aspirations in connection with Nikolai Semenov's 1956 Nobel Prize in Chemistry.[45] The leadership of the Division of the Physical–Mathematical Sciences of the Soviet Academy of Sciences adopted Resolution No. 19, of 1 November 1955, which said, amongst other things:

> The divisional leadership does not find it advisable to nominate Soviet scientists for the Nobel Prize since this prize cannot be considered international as demonstrated by the lack of Nobel awards to outstanding individuals of science and culture of our country (D. I. Mendeleev, L. N. Tolstoi, A. P. Chekhov, M. Gorkii).

Well-known scientists, including Petr Kapitsa (P78) signed it, and a representative of the Communist Party was also present. The Soviet Academy was structured in divisions; the mathematicians and physicists shared the same division, whereas the chemists and biologists each had their own, among many other divisions.

The timing of the resolution was significant, since it was expected in Moscow in October 1955 that Nikolai Semenov would be receiving the Nobel Prize in Chemistry for that year. Instead, it was awarded to an American, Vincent du Vigneaud.

In fact, there had been no Soviet Nobel Prize up to that time. The two Russian Nobel Prizes, to Ivan Pavlov (Mo4) and Ilya Mechnikov (Mo8), were awarded during the time of the czars. A Russian writer, I. A. Bunin, had been given the Nobel Prize in Literature for 1933, but this made the situation worse, since he was an emigrant living in Paris.

In 1954 the secretary of the Biology Division of the Soviet Academy of Sciences, the biochemist A. Oparin, famous for his theory of the origin of life, attended a meeting in Stockholm. Upon his return to Moscow he reported to the president of the academy, A. N. Nesmeyanov, on the discussion he had in Stockholm with two members of the Nobel Committee for Chemistry. A. Tiselius (C48) and A. Fredga told Oparin that they would like to see Soviet scientists participate in the Nobel movement. They asked Oparin to discuss this with the Soviet authorities in Moscow. Nesmeyanov and Oparin told the Communist Party about Oparin's experience. Apparently these events and L. G. Sillén's activities (p. 18) influenced the decision about the chemistry prize in 1956, which Cyril N.

Hinshelwood of Oxford and Semenov shared 'for their researches into the mechanism of chemical reactions'.

It is not difficult to imagine that, back in October 1955, the Moscow forces against closer connections with the West must have rejoiced when the 1955 chemistry prize went to an American scientist rather than to Semenov. The physicists and mathematicians had little at stake at that time. The physicists did not think that there was any Soviet physicist among the potential nominees, and there is no Nobel Prize in mathematics. The physicists, though, may have been mistaken because the first Soviet Nobel Prize in physics was awarded very soon after, in 1958, to Pavel Cherenkov, Ilya Frank, and Igor Tamm 'for the discovery and the interpretation of the Cherenkov effect'. And it was not the first time that Soviet physicists had been considered for the Nobel Prize. When preparing the nominations for the physics prize of 1939, Niels Bohr considered a joint nomination of the American Ernest Lawrence and the Russian Petr Kapitsa but, in the end, dropped Kapitsa.[46] Lawrence received the Nobel Prize in 1939 and Kapitsa had to wait for his until 1978.

In December 1955, the Chemistry Division of the Soviet Academy decided not to nominate anybody for the 1956 Nobel Prize.[45] This was a safe decision, though, because the Nobel committee had already received the documentation supporting Semenov's nomination. The chemists' resolution was an exercise in political expediency.

That politics was involved is obvious from many episodes. Thus, for example, the secretary of the Chemistry Division of the Soviet Academy asked the Soviet Embassy in Stockholm in 1954 to reassure Professor Sillén that 'we consider participation in the contest for the Nobel Prize to be a great honor for any scientist'. Such a request could not have been transmitted without the support of the state security services.

Currently, every invitation to nominate candidates is accompanied by stern warnings to discourage any discussion of possible candidates in academies or other collective bodies. In the above story such bodies and their politics were deeply involved and the Nobel organization could not have been unaware of such involvement. On the contrary, they seemed to encourage it. Since materials in the Nobel Archives become available for research after 50 years, it might be possible to find out more about the circumstances of Semenov's nomination on the Swedish side in 2006.

There must have been a lot of discussion about the Nobel Prize in Moscow behind the scenes. The years between Stalin's death in 1953 and the twentieth party congress in 1956 were full of alternating, pendulum-like decisions. More information about this matter may only come out when all the documents of the Soviet Communist Party become available for research.

It is a minor, but telling, side issue that the children of Soviet laureates did not accompany their parents to the Stockholm celebrations.[47] As soon as Semenov's Nobel Prize became known, preparations for the trip to Stockholm began. The Semenovs had a son and a daughter. The Soviet officials informed the Semenovs at once that they could only take one of their children with them to Stockholm. The Semenovs chose their daughter. However, shortly before the trip Semenov was 'advised' that they should not take their daughter with them, either. The authorities even provided a reason for her withdrawal, which could be given if asked for in Stockholm. The Semenovs were supposed to say that Lyudmilla was unable to join her parents in Stockholm because of her college examinations.

Two years later, in 1958, Cherenkov was planning to go to Stockholm with his wife. The wives of Tamm and Frank were ill and they wanted to take their sons with them. The Soviet Embassy in Stockholm was duly requested to provide help in securing five formal suits for men, two of size 50 and one each of 46, 52, and 54. Four days before the trip, however, the Communist Party intervened and Tamm's and Frank's sons stayed home.

In 1964, along with the American Charles Townes, the Soviets Nicolai Basov and Aleksandr Prokhorov won the Nobel Prize in Physics 'for fundamental work in the field of quantum electronics, which has led to the construction of oscillators and amplifiers based on the maser–laser principle'. Mrs. Basov and Mrs. Prokhorov were to go, and the Prokhorovs also wanted to take their son, a student at Moscow University, with them to Stockholm. It did not happen.

There was a complication of a different sort in 1975. There were two Nobel Prizes that year to Soviet scientists. Leonid Kantorovich shared the economics prize with the American Tjalling Koopmans, and Sakharov received the peace prize. Not only could Sakharov not go to Oslo, but there was also a protesting statement by Soviet academicians against his prize. Kantorovich refused to sign the document and did not even raise the question of his son's trip to Stockholm.

The first Soviet scientist who succeeded in taking his family to Stockholm was Petr Kapitsa, who received half the physics prize in 1978. He was adamant in his determination and he succeeded. We will see in a later chapter (p. 115), too, that this was not the first time that Kapitsa had stood up to the Soviet authorities. His behavior was unique. Not long before the great purges of 1937 started in the Soviet Union, Ivan Pavlov (Mo4) had told Kapitsa, who had recently been detained in Moscow: 'You know, Pyotr Leonidovich, I am the only one here who says what he thinks and when I die you must continue to do this. Speaking out is essential for our country, which I have come to love especially strongly in these difficult times'.[48] Pavlov died in 1936, at the age of 87.

Benevolent nationalism

It sometimes happens that nationalism, however benevolent, creeps into a nomination. In an unsuccessful nomination for the Nobel Prize in Chemistry for 1939, N. R. Dhar of Allahabad, then in India, sent a letter to the Nobel Committee[49] proposing the Frenchman Georges Urbain for his excellent researches on rare earth chemistry. Following a description of Urbain's achievements in chemistry, Dhar notes that Urbain is the doyen of French chemistry and a real gentleman. He would find it fitting if the land of Berzelius should honor the great land of Lavoisier. He also notes that the two great nations, the Scandinavian [sic] and the French have worked shoulder to shoulder in the development of rare earth chemistry and he sees here a great opportunity for one nation to recognize someone of the sister nation. In Dhar's time his expressions raised fewer eyebrows than they would today. In the presentation speech at the 1926 award ceremonies the French physics laureate Jean Perrin was greeted 'as a representative of the glorious science of France.'[50]

Various surveys are of interest in learning about polls that may not be free of national sentiments. The British *Chemistry & Industry* conducted a poll in 1999 for the Scientist of the Millennium. Only four living scientists made the list, with Frederick Sanger in the leading position. Sanger is the only person to have received two Nobel Prizes in chemistry, in 1958 and 1980. A few years before, the American Chemical Society's *Chemical & Engineering News* conducted a poll for its 75th anniversary to establish an international list of 75 people who, during the past 75 years, had contributed most to chemical enterprise. Curiously, the British Sanger did not make that list of 75, even though the poll targeted chemists only.

It is a delicate question whether the authorities in any country are willing to actively help their scientists receive a Nobel Prize. Lobbying may be done with taste and honest means, though nobody is eager to go on record about what activities may be carried out in this respect. Recently, a noted science historian in Budapest was invited to prepare a report on whether it was feasible to facilitate creating another Nobel laureate in Hungary. The analysis of Nobel histories concluded that, although the result could not be guaranteed, it was worth trying.[51] Gösta Ekspong[52] was recently asked in China for advice about how China, that is, scientists in the People's Republic rather than Chinese individuals abroad, could win a Nobel Prize. Ekspong suggested going to the schools, finding the clever children, cultivating them, giving them support, creating the right conditions for research at the appropriate time, and letting them think freely. He sensed, however, that this was not the party line.

3
Who wins Nobel Prizes?

I once made a frivolous comparison in an after-dinner talk between winning the Nobel Prize and becoming Miss America. I had read about a company in Texas that specialized in preparing young ladies for winning the Miss America title. There were many specific features of their preparation, like plastic surgery, that are not relevant for our comparison. However, the company carefully studied the prevailing public opinion, then adjusted the candidate's characteristics to best approximate it. If they found it necessary, they would even launch a campaign to sway public preferences towards the characteristics of their leading candidate. Although it would be outrageous to suggest anything similar in the quest for the Nobel Prize, the Swedish judges cannot ignore public opinion and the expectations of the scientific community, whether consciously or unconsciously. In spite of the independence of the Nobel institution, the judges care deeply about the approval of the international scientific community. Conversely, public relations may promote a discovery and may make the scientist and his work better known; it also helps if the scientist is a good performer.

Of all the people working in the science fields that correspond to the Nobel categories, a maximum of nine may become Nobel laureates annually. The odds are so low that people might just as well forget about it. In reality, however, exactly because of Nobel's intentions of rewarding great discoveries rather than great scientists, a much broader constituency feels potentially eligible for the prize than would be the case if the pool for consideration was determined by greatness. Theoretically, any scientist may strike gold, may make a serendipitous discovery, and this makes the prize look more reachable.

There is little in common between winning the Miss America title and the Nobel Prize, except the after-dinner speech. One of the most striking differences is that, whereas the Miss America candidates un-abashedly declare their intentions, would-be Nobel laureates seldom do, openly, that is. Eugene Garfield, who instituted an information revolution in scientific research,[1] noted that while the need for recognition is a universal motivation, most scientists find this fact of life embarrassing.[2] When Frederick Sanger (C58, C80) congratulated his colleague, John Walker (C97), on winning the prize, he added, ominously, 'Win another one!' This must have been a turn-off for Walker, who soon accepted a high administrative position. Walker thinks that one's personal ambitions should be kept to oneself, and he has sent away people who wanted to talk to him about his possible Nobel Prize. He has seen people over the years, brash young people, talking about their doing a Nobel Prize-winning ex-periment, and many clearly became disappointed when they felt they mer-ited the prize and did not win it. He noticed that 'It drives people crazy.'[3]

Although many people aspire to win the Nobel Prize, it is *not* customary to admit it, let alone to set up a training program for it. That would be sacrilege award often tell about their surprise at the announce-ment of their award. On the other hand, there is often information about an early teacher, a grandmother, or a girlfriend who had predicted their prize decades before. Of course, nobody cares to remember all those pre-dictions for people who never win.

Greatness in question

Scientists often, and the general public invariably, think that the Nobel laureates are supposed to be the greatest scientists. However, the Nobel Prize is often awarded to someone who is not counted among the 'great scientists'. The prize is supposed to be given for a specific discovery rather than for a lifetime achievement, and specific discoveries are not neces-sarily made by the most knowledgeable scientists of the highest authority. René Dubos said that

> many important discoveries have been made by men of very
> ordinary talents, simply because chance had made them, at the
> proper time and in the proper place and circumstances, recipients
> of a body of doctrines, facts and techniques that rendered almost
> inevitable the recognition of an important phenomenon. It is
> surprising that some historian has not taken malicious pleasure in
> writing an anthology of 'one discovery' scientists. . . . Science now

and then selects insignificant standard bearers to display its banners.[4]

Respectable scientists criticized Kary Mullis's (C93) prize for the polymerase chain reaction, not because they did not think the discovery important but because they did not think Mullis worthy of it. Referring to the potential applications of the polymerase chain reaction, which ranged from clinical diagnostics to criminology, the famous American TV journalist Ted Koppel said, on ABC's 'Nightline',

> Take all MVPs[5] from professional baseball, basketball, and football. Throw in a dozen favorite movie stars and a half dozen rock stars for good measure, add all the television anchor people now on the air, and collectively, we have not affected the current good or the future welfare of mankind as much as Kary Mullis.[6]

Mullis seems to have been carried away by his celebrity status and has since made anti-science statements. His Nobel Prize, however, was tailor-made for demonstrating what Alfred Nobel must have meant in his *Will*.

The 1993 physics prize went to Russel Hulse and Joseph Taylor, both of Princeton University, for the discovery of the double pulsar. The discovery was made by Hulse, a PhD student who eventually left the field and became a plasma physicist. He has been good in his new field, but not one of the top persons. Then this plasma physicist was asked to come to Stockholm, many years after the discovery, to receive the Nobel Prize in Physics for his part in discovering the double pulsar. The co-recipient was his former adviser, who had given him the job of looking for what then became the double pulsar and who remained a leading figure in the field. Many questioned whether Hulse should have been given the prize, although adding a person to the Nobel roster is a lesser responsibility than omitting someone. The award to Hulse was completely legitimate in view of Nobel's *Will*.

The discovery of the Raman effect (cf. p. 29) was made in the close partnership of C. V. Raman (P30) and his associate K. S. Krishnan. The effect should have been called the Raman–Krishnan effect.[7] Apparently, more than just the discovery was taken into account when the judgement was made for the Nobel Prize. Raman himself wrote: 'If the Nobel award for Physics made in 1930 had been based on the record of the year 1928 alone, instead of the entire work on the scattering of light done at Calcutta from 1921 onwards, Krishnan would in justice have come in for a share of the Prize.'[7a] The effect, under the name of 'combination scattering', was also discovered in 1928 by the Russian scientists G. S. Landsberg and L. I. Mandelstam. Their omission from the Nobel Prize has been discussed.[7b]

When someone receives the Nobel Prize their name is entered in the book of immortals. However, this is not the Nobel committees' doing, it was not even Alfred Nobel's doing. Nobel did not mean to single out the greatest scientists, the greatest minds, rather, he meant to single out the persons that 'have conferred the greatest benefit on mankind'. A lucky hand may qualify for this distinction just as well as the most knowledge-able person, because the criterion is not in the person, it is in the discov-ery. Anders Bárány has strong views on this: 'Considering Nobel's *Will* the Taylor-Hulse award in 1993 was true to the letter and spirit of the Nobel Prize and the omission of Jocelyn Bell from the pulsar prize in 1974 was a grave violation of it.'[8] We shall discuss Jocelyn Bell's exclusion in Chapter 12. Bárány continues:

> This is a tremendous change in the approach to the Nobel Prizes. In earlier times it was strongly advocated that a person receiving the Nobel Prize should . . . be a marvelous personality, should be some sort of a 'super-professor,' having lots of students and so on. For a long time, even in the 1960s and early 1970s, it was the prevalent opinion. By 1993, certainly this had changed, and there was a return to the original Nobel ideas in that the importance of the personality, the importance of a big school credited to the person, the position of the person in the scientific community had diminished and the actual discovery became the focus of attention almost exclusively. The prize could be given even to a complete fool who merely stumbled into something important.[8]

If the Nobel Prize were given for lifetime achievements, scientists like Albert Einstein could have won on that basis, but this was not the case. Einstein did receive the prize, but not for his best known contributions to physics, the special and general theories of relativity. He was awarded the prize 'for his services to theoretical physics, and especially for his dis-covery of the law of the photoelectric effect'. Although Einstein did not attend the award ceremony, there is an Einstein lecture printed in the *Nobel Lectures* volumes, which he delivered in Gothenburg in 1923, and it is on the theory of relativity.[9] The physics Nobel committee's caution in more openly embracing Einstein's theory of relativity is symptomatic of the fact that, at the time, Swedish science was not very strong in theoretical physics. When Oskar Klein was appointed professor of mechanics and mathematical physics in Stockholm in 1930, his good friend Wolfgang Pauli (P45) welcomed this because 'Until now there was practically almost no theoretical physics in Sweden'.[10]

Ernest Rutherford, another of the greatest physicists of all time, never received the Nobel Prize for his nuclear model of the atom. He was awarded the prize in chemistry for 1908 'for his investigations into the disintegration of the elements, and the chemistry of radioactive substances'. In 1923, the motivation for *not* awarding Rutherford another Nobel Prize was spelled out by the physics committee:

> It is understood that Sir Ernest's merits are so great and generally acknowledged that his standing outside Sweden would not markedly increase by the award of a new Nobel Prize, nor would it markedly increase his possibilities for research, as these already are as great as possible after he received the most prestigious research position in the field of physics in the British Commonwealth.[11]

It does happen that the Nobel Prize is given for a lifetime achievement without explicitly stating that. Eugene Wigner's Nobel Prize in Physics for 1963 was 'for his contribution to the theory of the atomic nucleus and the elementary particles, particularly through the discovery and application of fundamental symmetry principles'. He received half of the physics prize and many wished that he had received it alone. Gerhard Herzberg's (C71) Nobel Prize is also one of those that honored a lifetime achievement rather than a specific discovery.

Hans Bethe (P67) received the Nobel Prize for his work on the theory of nuclear reactions and in particular on energy production in stars. Oskar Klein[12] of the physics committee gave the presentation speech at the award ceremony. He supposed that Bethe was 'astonished that among your many contributions to physics, several of which have been proposed for the Nobel Prize, we have chosen one which contains less fundamental physics than many of the others and which has taken only a short part of your long time in science'. Klein stressed that their choice was in agreement with the rules of the Nobel Prize.

From time to time there appears to be a dilemma in Nobel judgements between awarding a specific discovery and singling out a great scientist. There are, then, other problems. Lars Ernster, who was an active Nobel committee member for many years, himself found some of the earlier Nobel decisions puzzling.[13] On some occasions the discoverers were given the award for something other than their most important findings. Otto Warburg (M31) received the prize 'for his discovery of the nature and mode of action of the respiratory enzyme', and Fritz Lipmann (M53) for his discovery of co-enzyme A. However, what was most important in Warburg's work was that he described ATP's role in glycolysis, and in Lipmann's that he determined that ATP was the high-energy phosphate. ATP is adenosine triphosphate and Karl Lohman discovered it in Germany in 1929. It is the universal energy currency for living cells.[14] Lohman did not receive the prize because he just identified it as a phosphate in muscle fibers. Soon after, Vladimir Engelhardt in Russia discovered oxidative phosphorylation in respiration. He did not receive the prize, either. The discoveries of Warburg and Lipmann were crucial for understanding ATP and bioenergetics, and they built upon Lohman's and Engelhardt's discoveries. In hindsight the works of all four were fundamental in creating the new science of bioenergetics. The impression is that Warburg and Lipmann were selected for the Nobel Prize because they were great scientists and it was of secondary importance for which of their achievements they were cited. Had they been cited directly for their greatest discoveries, their ATP-related work, the omission of Lohman and Engelhardt from the Nobel roster would have appeared more conspicuous.

Haves and have-nots

While it is not very plausible to give a description of a typical Nobel laureate, a few comparisons between laureates and non-laureates may help us discern some characteristic traits of their personalities and careers.

Comparison I William Astbury (1898–1961) of the University of Leeds, and Linus Pauling (C54, Peace 1962). According to J. D. Bernal,[15] the history of molecular biology will show that the work of Astbury, from its beginnings in 1926, was the main line of progress in this field. It started with the appreciation of the α fold, which is called the α-helix today.

Later, Linus Pauling discovered the α-helix. Bernal was intrigued by the difference between the two great scientists. His concern was not with why Astbury did not get the Nobel Prize, but with why Pauling did. He wanted to understand why Astbury did not make more discoveries, as Pauling did, and why he did not notice what others did on the basis of his own original work. The two men were of the same generation; both had similar training as chemists. Astbury was not backward in disseminating his findings, publishing his results widely. However, Pauling incorporated what he learned about the quantum theory into chemistry, along with a lot of factual information, thus creating a new structural chemistry, and this was his forte. Part of this was the metrical aspects of three-dimensional geometry. Pauling was shocked at how loosely X-ray crystallographers handled geometrical details in their structure elucidation. The most conspicuous example was how Perutz (C62), Kendrew (C62), and Bragg (P15) ignored the planarity of the peptide bond in their numerous proposed models for proteins, which all proved to be wrong.[16] Pauling may have appeared to be flamboyant in presenting his results but he was meticulous, primarily through the efforts of his Cal Tech associate, Robert Corey (1897–1971), in gathering his data before he came up with his revolutionary models. Astbury was more isolated than Pauling; his pioneering spirit may have been a disadvantage to him in that he found solutions long before others but also clung to them long after they had been abandoned by others. Linus Pauling developed a similar habit but it came much later in life. His resistance to the new idea of quasicrystals versus the well-established dogmas of crystallography shows a similarity with Astbury's behavior. Many of Astbury's achievements were so important that they have become commonplace and are around without being associated with Astbury's name anymore.[15]

Bernal recognized the importance of Pauling's contribution: 'He knew his atoms and their various states and binding conditions so well that he was prepared to break with what are after all only conventions—such as the regularities of classical crystallography—if they could not be fitted into these regularities.'[17] (See also on the α-helix discovery, p. 93.)

Strong vanity is ascribed to Linus Pauling. When Max Perutz provided unequivocal evidence for the correctness of Pauling's α-helix model, he expected Pauling to be pleased. Instead, Perutz writes, 'he attacked me furiously, because he could not bear the idea that someone else had thought of a test for the α-helix of which he had not thought himself.'[18]

Comparison II Walter Gilbert (C80) and Eugene Sverdlov. Benno Müller-Hill draws an interesting parallel between them in his book, *The* lac *Operon: A Short History of a Genetic Paradigm*.[19] Gilbert received the Nobel Prize for work on DNA sequencing, known as the Maxam–Gilbert method; at the time Allan Maxam was Gilbert's technician. There was a Russian side of the story in that Eugene Sverdlov and his co-workers[20] had published a similar idea. This work had appeared in English in an international periodical and was thus accessible. Besides, Sverdlov had personally given a reprint of his paper to Gilbert at a meeting in Kiev in 1975. Gilbert never referred to Sverdlov's work in his publications. When Müller-Hill, Gilbert's former postdoctoral associate, had described this and had asked Gilbert for comment prior to the publication of the book, Gilbert had no comment. Müller-Hill compares the two scientific careers. Gilbert grew up in an environment most conducive to a career in science, while Sverdlov was handicapped by many of the disadvantages the Soviet system could inflict upon him. Sverdlov did not continue his research in sequencing of the DNA because his director of the institute of the Soviet Academy of Sciences did not believe in its feasibility. That director was Yuri Ovchinnikov,[21] who had a meteoric career in Soviet science and was much admired in the West, but Müller-Hill labels him 'the great tyrant and charming actor'. Today Sverdlov is the director of the Institute of Molecular Genetics of the Russian Academy of Sciences in Moscow.

Comparison III Murray Gell-Mann (P69) of Cal Tech and Yuval Ne'eman[22] of Tel Aviv University. Gell-Mann was awarded the prize 'for his contributions and discoveries concerning the classification of elementary particles and their interactions'. In his presentation speech the Swedish academician[23] described Gell-Mann's achievements in which the symmetry used to classify the particles, 'The Eight-fold Way', occupied a central place. The Israeli physicist, Yuval Ne'eman had published the same suggestion independently, alas without giving it a name. This was not the only occasion when Gell-Mann and Ne'eman ran along parallel paths in physics. Both of them had predicted elementary particles, on the basis of their classification, that were eventually identified. This was similar to the well-known case in which Dmitri Mendeleev predicted unknown elements for his periodic table, providing its ultimate test.

There are scientists who are more predisposed for recognition and others who are less so. Ne'eman[24] had considerable handicap. He was a latecomer to physics and unknown when he burst into this close-knit, small world. To many he was an amateur, an Israeli army colonel and engineer who did physics in his spare time. Ne'eman had loved physics from his high school days and in 1958 he joined Abdus Salam (P79) at Imperial College in London as a graduate student. By then he was 33 and the defense attaché at the Israeli Embassy in London. He had devoted his earlier years to fighting for Israel's independence. From the spring of 1960 he could finally concentrate on physics. In his first studies, it turned out, he duplicated already existing discoveries, but this encouraged rather than disappointed him. Then he discovered the applicability of the SU(3) Lie symmetry group to the systematization of the elementary particles, essentially the same thing to which Gell-Mann later gave the fancy name 'The Eight-fold Way'. Ne'eman and Gell-Mann also overlapped in other research, including the conception of quarks. At some point Gell-Mann invited Ne'eman to Cal Tech and the two scientists found a lot in common beside their love for physics. In their childhood they both liked to classify things, minerals, languages, the Chinese dynasties, the kings of Europe. They were both Jewish, though this meant very different things to them.

The systematization of the elementary particles and the prediction of unknown particles that were eventually identified brought fame and recognition to the discoverers. But it was Gell-Mann who stayed in the limelight with Ne'eman's contributions known only to the specialists. Gell-Mann's biographer writes: 'Soon people were talking about Gell-Mann as the new Mendeleev. (If they were scrupulous with their citations, they conferred the honor on Ne'eman as well.)'[25] One can only speculate about a possible career in science for Ne'eman, had he led a life typical of his generation in the United States. He might have grown up in a poor immigrant family, and he might have had it instilled into him that his only way of ascent, if he wanted to succeed, would be studying and doing science at least twice as hard as his non-Jewish peers. There would have been no diversion of a brilliant military career and so on and so forth.

The stories about Murray Gell-Mann are consistent. To quote a former pupil of his:

he was about to enter a lecture class and Gell-Mann arrived at the
door to give the class. My friend was about to open the door but
was stopped by Murray, who said, 'Wait!' There was a storm
raging outside the building, and at the appearance of a particularly
violent flash of lightning, he said, 'Now!'—and entered the class
accompanied by a duly impressive peal of thunder.[26]

According to Sheldon Glashow (P79) Murray Gell-Mann 'knows almost
everything about almost everything, and he is not averse to letting you
know that he does and you don't.'[27]

Strong personalities are not an exception among Nobel laureates.
This is how Gilbert Stork of Columbia University remembered former
Harvard colleague Robert Woodward (C65):

Woodward was an extraordinary individual, but he had an implicit
belief that if he did not produce or suggest something, it had no
particular importance. That is to say, until the Pope had given his
approval to something, it didn't exist. So he didn't have to give
too much credit to previous workers because, by definition, until
Woodward said it was all right, it was not.[28]

Comparison IV Erwin Chargaff (b. 1905) and James Watson (M62).
Watson was, of course, the co-recipient of the Nobel Prize in Physiology
or Medicine, together with Francis Crick and Maurice Wilkins, for the
double helix structure of deoxyribonucleic acid, DNA. When the discovery happened, in the spring of 1953, Watson was 25. Even though
not a child prodigy, he had done his studies fast and was on a postdoctoral
fellowship in Cambridge. Watson and Crick made their discovery by
model building. They collected all available information about the DNA
molecules and tried to build up a three-dimensional molecular structure,
like an erector model, that would conform to all the data in their posses-
sion. They knew about the sugar and phosphate parts of the four nucleotides, each of which contained a different base, adenine, cytosine, guanine,
or thymine. Of these, cytosine and thymine belonged to the same class, the
pyrimidines, and adenine and guanine to the purines. A crucial piece of
information that went into their model building was the so-called base
complementarity, established a few years before by Erwin Chargaff, the
Austrian-born biochemistry professor of Columbia University. Chargaff,
who was in his mid-forties at the time of this discovery and who had

had a tremendous classical education in all the foundations of his science and beyond, had switched to DNA studies after Avery and his associates had determined that DNA was the substance of heredity (see, p. 225). He was one of the few scientists who recognized the importance of Avery's discovery at once. Base complementarity means that the amount of guanine is the same as the amount of cytosine and the amount of adenine is the same as the amount of thymine.[29] This relationship was found to be valid in DNA of different origins.[30] Base complementarity introduced severe restrictions into the DNA model building, providing one of the keys to the solution. Although Chargaff had published his observations, Watson and Crick were not very meticulous in following the literature and they learned about them from Chargaff himself on his visit to Cambridge. We know from Watson's *The Double Helix*[31] that Chargaff did not find his younger colleagues adequately versed about the bases; they appeared ignorant of their structural formulas. Indeed, Watson had had a deficient preparation in chemistry (Crick was a physicist) and would learn it after the double helix discovery, but he did do so, as witnessed by his superb text, *The Molecular Biology of the Gene*.[32] When Watson and Crick discovered the double helix model, its beauty and biological implications convinced them and most others that it had to be right. They dispatched a barely one-page letter to *Nature*,[33] and revolutionized science for the rest of the century and beyond. They could have found reasons to delay the publication, they could have devised further tests, asked for more data from others; nobody could have blamed them for some hesitation. In contrast, Chargaff's description of base complementarity is pregnant with skepticism and hesitation as he finds it 'noteworthy—whether this is more than accidental, cannot yet be said—that in all desoxypentose nucleic acids examined thus far the molar ratios of total purines to total pyrimidines, and also of adenine to thymine and of guanine to cytosine, were not far from one.'[29]

Chargaff inserted this paragraph at proof stage during the process of the publication of his discovery.[34] This paragraph, although formulating the observed regularity, reflects Chargaff's hesitation, as if he were trying to fight off his discovery. Three decades later, Chargaff explained the origin of his skepticism: 'For a long time I felt a great reluctance to accept such regularities, since it had been impressed on me that our search for harmony, for an easily perceived and pleasing harmony, could only serve to distort or gloss over the intricacies of nature.'[35]

In hindsight it is easy for us to judge that Chargaff did not recognize what Watson and Crick did, namely, that there is a time for the general picture and there is a time for the details. Around 1950 the big question about DNA was a general one, pertaining to the relationship between its structure and function. Witnessing the triumphal march of the double helix model, and remembering Oswald Avery's contribution (as well as, undoubtedly, his own), to the DNA story, Chargaff bitterly remarked: 'What counts . . . in science is to be not so much the first as the last.'[36] Chargaff describes how he sees himself:

> In some ways I was the wrong man to make these discoveries: imaginative rather than analytical; apocalyptic rather than dogmatic; brought up to despise publicity; uncomfortable in scientific gatherings; fleeing all contacts; always happier with my youngers than with my betters; more afraid of an absurd world than trying to understand it; but ever conscious, day and night, that there is more to see, more to say than I can say, and even more to be silent about.[37]

This is hardly the description of our typical Nobel laureate, although I am not implying that Chargaff did not receive the Nobel Prize because of these traits in his personality. In fact, he received other prestigious awards and distinctions and every biochemistry text discusses his observations of base complementarity and species-specificity of DNAs. As for Watson, not only did he win the Nobel Prize, his straightforward march to a crucial problem of molecular biology and its relentless solution, learning the necessary knowledge and information as the need arose on the way, was emulated in numerous success stories during the second half of the twentieth century.

Comparison V   Stanley Prusiner (M97) of the University of California, San Francisco, and Charles Weissmann (1931),[38] formerly of the University of Zurich. In 1997, Stanley Prusiner was awarded the Nobel Prize in Physiology or Medicine 'for his discovery of prions—a new biological principle of infection'. According to some, Weissmann should have shared it with him. In the mid-1960s, a mathematician, John Griffith, proposed the idea that a protein could be an infectious agent. Prusiner took up this idea, coined the name prion for it (cf. p. 186), and provided supporting evidence. The prion story has enormous importance because it relates to the rare Creutzfeldt–Jakob disease in humans and to the transmissible spongiform encephalopathies, which include the bovine spongiform encephalopathy (BSE), known also as 'mad cow disease', and scrapie, a similar disease of sheep.

Prusiner isolated the prion protein, PrP (later called PrP^{Sc}, referring to scrapie), from the brain of diseased animals. At first he thought that it was present only in infected brains, but later it was shown that the same protein was present in both healthy and diseased animals. Chemically they were the same, that is, they had the same amino acid sequence, but their spatial arrangement, the conformation, was different. Once the animal's brain is infected, the diseased form of PrP, that is, PrP^{Sc}, spreads, and nobody knows for sure how this replication happens. Provided there is no nucleic acid involved in this replication, Carleton Gajdusek's (M76)[39] suggestion is an attractive possibility. He likens the replication of the diseased protein to the mechanism of nucleation and crystallization.

Weissmann determined the sequence of PrP and, using the genetic code, he also determined the corresponding DNA sequence for it. Prusiner and Weissmann joined forces at some point and detected this DNA sequence in scrapie-infected animals. To their great surprise, they detected it in healthy animals as well. In further experiments, Weissmann and his group knocked out the PrP gene and demonstrated that even the healthy PrP was not necessary for normal and healthy animals. Apparently this gene was conserved during evolution in all vertebrates, was present in the brain and other organs already in the early embryonic development, and yet it could be eliminated without damage. The gene-knockout experiment could have falsified the prion–only hypothesis but it spectacularly failed to do so and, instead, provided convincing evidence for it.[40] This experiment could have been the strongest claim for Weissmann's inclusion in the Nobel Prize for prion. The decisive experiment for proving the protein-only hypothesis is still missing. This would involve taking the normal PrP and converting it under controlled conditions into PrP^{Sc}.

Prusiner was very determined to get the Nobel Prize. If there was strong lobbying on his part, it must have been the kind that the Swedish jurors did not resent and may have found useful in turning their attention to important discoveries. Weissmann, on the other hand, did nothing to promote his candidacy. He thought that if there were merit in what he had done it would be recognized in Stockholm without his asking people to submit nominations on his behalf. He did receive a number of prestigious awards, but alas not the ultimate one. His recognition materialized not only in his many awards: he was also as appointed to chair the European Commission Group on Bovine Spongiform Encephalopathy in 1996. His achievements have crossed several areas of molecular biology and genetics. During the 1980s he was best known for the first synthesis of biologically active interferon. After getting his MD, he studied with Paul Karrer (C37) and took his PhD degree. He then spent seven years with Severo Ochoa (M59) in New York. From 1967, until his retirement, he directed the Institute of Molecular Biology at the University of Zurich. He is now senior research scientist of the Neurogenetics Unit at Imperial College in London.

Drilling and digging

Lars Ernster classified scientists into two categories, the drilling type and the digging type.[13] The drilling type pursues the same project throughout an entire career and may or may not make an important discovery. The digging type changes from topic to topic and in a lucky case may make one or several important discoveries. The difference exists not so much in research strategies as in personalities. Using Ernster's terminology, Prusiner is a driller and Weissmann a digger. Prusiner staked his whole career on the prion project. One of the greatest risks, if not the greatest, he took was the introduction of the name prion. It pledged him to the prion-only hypothesis early on. On the other hand, it forever associated him with it.

The question arises of whether a drilling or a digging type has a better chance of winning the Nobel Prize? The arguments can go both ways. As the prize is, at least theoretically, given for a discovery rather than for a lifetime performance, the drilling type has a better chance of producing something of a breakthrough in one given area. On the other hand, the risk is greater in putting all one's eggs in the same basket. The digging type increases the probability of discovering something while moving from area to area. On occasion, even though lacking a specific striking discovery his findings may be lumped together and awarded. Aaron Klug is a typical digging case and his Nobel Prize in Chemistry for 1982 was 'for his development of crystallographic electron microscopy and his structural elucidation of biologically important nucleic acid-protein complexes'. This contrasts well with the much more specific citation for the Perutz–Kendrew prize in 1962: 'for their studies of the structures of globular proteins', which taken separately meant exactly myoglobin for Kendrew and hemoglobin for Perutz. Max Perutz is the example par excellence of the drilling type. He took up the problem of the hemoglobin structure, at someone else's suggestion, in 1936, and has stayed with it for his entire career, which still continues in his late eighties. Even as Perutz attended the Nobel ceremonies in Stockholm in 1962, more and better X-ray diffraction data were being collected in his laboratory and plans were drawn for the improvement of the hemoglobin structure.

Most Nobel laureates received the honor late in their lives, although their discoveries have usually, though not always, been made in their youth. One of the rare exceptions was William Henry Bragg (P15), who came to research relatively late. He had a chair in mathematics and physics in Adelaide, Australia after he had studied in Cambridge. He started doing research at the age of 42, in the course of the preparation for a lecture on the ionization of gases, and initiated a correspondence about his results with Ernest Rutherford (Co8). Five years later he was elected as a Fellow of the Royal Society and given a chair in physics at the University of Leeds. His son was studying in Cambridge and in 1912 he learned about the exciting news of the Munich experiments on X-ray diffraction on a crystal set given to him by his father. The young graduate student and the latecomer father then initiated X-ray crystallography.[41]

Max Planck (P18) worked out his quantum theory, probably the most revolutionary theory of physics in the twentieth century, when he was 42 years old, in 1900. His previous activities had not suggested that he might one day come up with such a breakthrough. It did not achieve wide acceptance easily and quickly, either. It was only in 1919 that he was awarded the 1918 Nobel Prize for it.

Charles Pedersen (C87) never did doctoral work, but spent a successful career as a research associate at Du Pont in Wilmington, Delaware. He was 57 years old when he embarked on the project that led to his Nobel Prize in Chemistry for 1987. In some serendipitous experiments he made some organic molecules that were readily capable of forming complexes with inorganic salts. Incidentally, there have been a few other Nobel laureates who did not have a doctorate, such as Gertrude Elion (M88) and James Black (M88). This bothered Elion throughout her life but apparently Black did not care.

In his nomination for the Nobel Prize in Physics in 1948, J. D. Bernal attached a long and detailed factual account of P. M. S. Blackett's achievements.[42] Although he is aware of the fact that personal traits are not strictly relevant for getting the prize, he notes that he admires and respects Blackett's scientific ability and calls him a worthy successor to Rutherford at the Cavendish Laboratory in Cambridge. According to Bernal, among the experimental physicists alive, Blackett is one of the greatest, even though no spectacular discoveries have fallen to his lot. This is almost a double jeopardy since greatness is not a consideration for the Nobel Prize whereas a great discovery is. Bernal adds that his protégé, by his character and ability, ranks among the distinguished men who have in the past received the prize. This is also a flawed argument, but apparently Bernal knew better because Blackett did receive the Nobel Prize in Physics.

In 1947, Harold Urey (C34) prepared a detailed analysis[43] of the chemical sciences for considering possible Nobel Prizes. He raised the question of the advisability of a second Nobel Prize and reviewed the work done on the atomic bomb, mentioning Fermi, Szilard, and Hahn. Of the three, both Hahn and Fermi had already received Nobel Prizes, although in the case of Fermi it was for different work. In the work on the atomic bomb it was he, more than anybody else, who drove the project of the chain reaction to a successful conclusion, although originally Szilard had thought of the idea and patented it. Urey believed that generally it is not a good idea to award two Nobel Prizes to one person, discarding even himself, but made an exception for Fermi. The second line of work to be considered for the Nobel Prize, according to Urey, should have been the discovery and separation of plutonium. For this work he proposed Glenn Seaborg (C51) and Joseph Kennedy, and noted that Kennedy's work had received much less publicity than Seaborg's. Another possibility for the award may have been the work on the separation of uranium isotopes, but here Urey issued a caveat, pointing out that some parts of the atomic bomb project had received more emphasis than others, and he advised waiting a couple of years before making decisions concerning these works. In two subsequent letters in April 1947.[44] Urey made two strong points about possible Nobel Prizes for works arising from the atomic bomb project. One was that no prize should be awarded for work that was kept secret. The other point was that work related to atomic energy should be rewarded only when peaceful applications had arisen, in keeping with Alfred Nobel's *Will*. By then Otto Hahn (C44) had received his Nobel Prize for nuclear fission. This prize was announced in November 1945, just a few months after Hiroshima and Nagasaki. Hahn received the news about his prize while in detention with other prominent German scientists, 'Hitler's Uranium Club', in England.[45]

'Third persons'

Some Nobel laureates are the 'third person' who happens to be positioned for the prize under fortunate circumstances and who is admissible because the allotment of the three-person rule makes it possible. We will cite two examples of the 'third person'.

Robert Curl shared the Nobel Prize in Chemistry for 1996 with Harold Kroto and Richard Smalley 'for their discovery of fullerenes'. Before the fullerene discovery nothing pointed to a possible Nobel Prize in his distinguished career. He was professor at Rice University, but he did not make it into the National Academy of Sciences. Then came the fateful week in September 1985, the discovery of the third (graphite and diamond being the others) modification of carbon, C_{60}, buckminsterfullerene. This discovery catapulted Curl into a Nobel Prize-winning position. Had the maximum possible number of awardees been two, only the principal players, Kroto and Smalley, would have shared the award. Had the maximum number of awardees been four, the additional two would have been Donald Huffman and Wolfgang Krätschmer (see, p. 242). They produced buckminsterfullerene in measurable quantities for the first time in 1990 and thus made it possible to ascertain its structure and carry out other experiments on the substance. It was, however, virtually impossible to delineate their contributions and prefer one to the other. Thus, there was an available third slot and it went to Curl, and in this the Nobel Committee acted consistently. Three senior scientists participated in the project at Rice University in which they made the observation on C_{60} and suggested a structure for it. It is important that all three professors were among the authors of the paper reporting the work.[46] There were also several graduate student participants,[47] but only the three senior scientists were included in the prize.

Curl[48] meditated on his role in the discovery for years. He noticed, amidst all the publicity that was going around from the time of the discovery to the time of the Nobel award, that 'my name was almost never mentioned. It was always Kroto and Smalley. In popular magazines especially, when fullerenes were mentioned, I was very seldom mentioned. Harry in Europe and Rick in this country were doing most of the interviews with the popular press.' Curl worried that he might be left out. However, he trusted the Royal Swedish Academy not to base its decision on the popular press and that it would appreciate that: 'I saw my job to be the one concerned that we weren't screwing up, that it wasn't all wrong. . . . I've had a long career of being the designated worrier, and I think my colleagues liked having a professional on the job.' Lest we underestimate Curl's input, such a contribution in the Perutz–Kendrew–Bragg attempt to solve protein structures could have saved them from a major embarrassment (p. 54).

Robert Huber's (C88) prize provides another example of the 'third person' notion. Huber was a division leader at the Max Planck Institute for Biochemistry in Martinsried, near Munich, Germany, an expert in protein crystallography who had built up one of the world centers in his field. At one point his laboratory was so prolific that it probably solved more protein structures than any other lab in the world. In addition he had introduced a series of innovations in protein crystallography.[49] Yet there was no reason to anticipate a Nobel Prize for him in protein crystallography; it was an area of research that had been pioneered in Cambridge, whose pre-eminence was not in question. Then Hartmut Michel (C88), in another division of the same institute, succeeded in crystallizing an exceptionally important protein, which is usually the bottle-neck in protein crystallography. He thus made it possible to determine the structure of a photosynthetic center, a long coveted goal in biomolecular chemistry. Thus the crystallization was a unique contribution and a tremendous breakthrough. At that point Michel could have asked any competent protein crystallographer to join forces with him to determine the three-dimensional structure. As it happened, Michel developed a harmonious and mutually enriching collaboration with Johann Deisenhofer (C88), who was one of Huber's former students and was still working in Huber's division.

The project to determine the three-dimensional structure of the photosynthetic reaction center was supposed to be the work of these two young researchers, work in which two other young colleagues, Otto Epp and Kunio Miki, also participated. Dieter Oestehelt, Michel's former mentor and current division head, kept out of the project. However, at some point Huber made it clear that he considered himself a participant. He thus levered himself into one of the winners' positions, as Michel and Deisenhofer's work headed for successful completion and it became clear that there was something truly big in the making.

There is no question that the structure elucidation of the photo-synthetic center deserved the Nobel Prize, and that the crucial step in this research was the crystallization of the membrane protein. The prize could have been awarded to Michel alone, or to Michel and Deisenhofer jointly. Yet another possibility was including Huber. If taken symmetrically, Oesterhelt might have been considered as Huber's counterpart, but he was not among the authors of the two crucial papers that carried five names altogether.[50] Huber's inclusion in the Nobel Prize was not incon-sistent with traditional Nobel practice of favoring senior participants. Considering the work in its totality, Oesterhelt might have been consid-ered also on the basis of the initial papers describing the crystallization of membrane proteins, which Michel and Oesterhelt published together in 1980. However, there was never a possibility of adding a fourth name. The Nobel citation read, 'for the determination of the three-dimensional struc-ture of a photosynthetic reaction center'. Thus it was very specific. From a superficial point of view there is nothing unusual about this Nobel Prize; a closer look, however, reveals an uncomfortable disproportion between Huber and Oesterhelt's attitudes and their rewards.[51, 52] However, before we pass a harsh judgement over Huber it should be realized that his indir-ect contribution to the prize-winning work must have been substantial. Whereas the first reaction-center crystal was X-rayed in 1981, Huber's involvement in protein crystallography had started in 1967.[53] Creating a protein crystallography laboratory, which has the capability of solving the largest structure to date and which is among the most prolific in the field, is no small accomplishment. Huber had built up what was needed for Deisenhofer's successful participation in the prize-winning project. Once Huber decided to count himself in, he provided valuable technical assist-ance in the X-ray experiments. At one point the technical personnel be-came unavailable for helping with the instruments and Huber substituted for them,[54] which might be a routine matter for an American professor but is rather extraordinary for a European one. His merits cannot be questioned; what can be questioned is whether or not he deserved the Nobel Prize for them. If Louis Pasteur is right about chance favoring the *prepared* mind in scientific discovery, it seems also true for the advantage of being prepared for the highest prize by being in the right place at the right time. As much as Curl's and Huber's careers were unexceptional from the point of view of the Nobel Prize, by virtue of those same careers they were well prepared to receive it.

We have already mentioned the prize in physiology or medicine 'for the discovery of insulin' awarded to Frederick Banting and John Macleod at Toronto University in 1923 (p. 7). Banting's associate, a young student named Charles Best, was overlooked and Macleod, who was not directly participating in the project but was head of the laboratory, was included. Peter Medawar[55] points out, in a balanced discussion, that Macleod had merits in the discovery. He provided Banting with the conditions for his research, and gave him technical advice and assistance. He was also instrumental in making a discovery that would benefit mankind rather than just the pharmaceutical companies. Medawar notes, 'All this deserves a place in history, but not a Nobel Prize; nor did it, on the other hand, deserve the contemptuous antipathy he received from Banting.' I suspect that had Best been included in the Nobel Prize and Macleod added as a third recipient, the uproar might have been muted. However, this is not what happened and Macleod's prize has entered Nobel history as an undeserved recognition.

Neither Curl's nor Huber's Nobel Prize raised eyebrows, except, perhaps within a small circle of insiders. No one was omitted among the immediate participants in the research projects concerned. In this there is a substantial difference between these cases and the insulin award. The 'third person' cases of Curl and Huber are very different, yet they have in common that their inclusion in the prize had about as much consequence as their omission might have had.

There may have been a missing 'third person' in the 1973 chemistry Nobel Prize, which was awarded to Ernst Fischer (C73) and Geoffrey Wilkinson (C73).[56] Robert Woodward (C65) of Harvard University, among others, wrote to the chemistry committee that 'the award of this year's Nobel Prize in Chemistry leaves me no choice but to let you know, most respectfully, that you have—inadvertently, I am sure—committed a grave injustice.' Woodward's claim to priority concerned the idea of the sandwich structure of ferrocene in 1952. Wilkinson also claimed that he had come to the idea independently. In fact, they published the novel structure jointly in 1952. At that time both were at Harvard, Woodward as full professor and Wilkinson as an independent assistant professor. Following their joint original work, Woodward left the field while Wilkinson dedicated himself to it, and so did Fischer in Munich. Woodward could have been included in the prize as a 'third person'. His own prize was in 1965 'for his outstanding achievements in the art of organic synthesis'. Thus, it was yet another of the general prizes rather than one for a specific discovery. In this sense the Swedish judges may have felt that Woodward's 1965 prize covered his future contributions to the field, not only the past ones. Second Nobel Prizes in the same field are rare; so far there has been one in physics and one in chemistry, each having been given for specific discoveries. The presentation speech[57] for the 1973 chemistry prize mentioned Woodward's name twice. Wilkinson[58] in his lecture did not mention Woodward at all, and he did not even quote their joint contribution. That the presenting Swedish academician mentioned Woodward by name was a departure from unwritten rules. In response to Woodward's protest, mentioned above, the chairman of the chemistry committee had informed him that 'it is customary not to mention co-workers and co-authors who are not sharing the prize. . . .'[56] It took a Nobel laureate's strong protest to make an exception, and, as a consequence, Woodward's contribution was given ample mention in the presentation speech.

An attractive extension of the circle of awardees happened with the Nobel Prize in Physiology or Medicine for 1954, which was given for growing the poliovirus in tissue cultures. Here the senior scientist, John Enders, might have been chosen as the only recipient. However, Enders went out of his way to make sure that his young associates, Frederick Robbins and Thomas Weller, were included.[59] The first paper about the successful cultivation of the poliovirus was published in 1949.[60] Soon after, a member of the Nobel Committee visited Enders. Robbins remembers vividly that Enders made it clear to the visitor that Weller and Robbins were very involved in the polio work. Robbins remarked, 'That Enders saw to it that his younger colleagues were included in the award speaks volumes for his character.'[61]

Traits

Although it is not necessarily the greatest scientists who receive the Nobel Prizes, laureates are nonetheless bombarded by the most diverse questions, as if they will know more about everything than others. Much humility is needed to resist the temptation to act as an oracle. This humility question was on the mind of the president of the Royal Swedish Academy of Sciences when she told the following story: a famous Nobel laureate is traveling with his wife in their car somewhere in the southern United States. They stop for gas. Suddenly she gets out of the car and the laureate is puzzled to see his wife hugging and kissing the gas station attendant. When the tank is filled and they drive away, the Nobel laureate asks his wife about the attendant. She tells him that he was her former sweetheart in her youth and she has not seen him for a long time. The Nobel laureate is not amused and tells her: 'It is embarrassing for me to see my wife hugging and kissing a gas station attendant. After all, I am a Nobel laureate.' To which she answers, 'Nobel laureate you are but had I married that guy, he would now be the Nobel laureate.'[62]

John Bardeen (P56, P72) won two Nobel Prizes in physics, and all who write about Bardeen remember him as a quiet and modest person. His hobby was golf, and many years after his second prize his long-time golf partner asked him, 'Just what is it you do for a living?' Apparently Bardeen had never worked the two Nobel Prizes into conversations with his golf partner.[63]

When Lawrence Bragg (P15) moved from Cambridge to London, he missed his garden and rented himself out through a newspaper advertisement, finding work as a part-time gardener. This went on, on a regular basis, until a visiting friend asked his employer what Sir Lawrence was doing in her garden.

In spite of these engaging examples, humility is not advocated for scientists in their research efforts. Peter Medawar (M60) gave this advice: 'Humility is not a state of mind conducive to the advancement of learning.'

George Olah (C94), and before him Georg von Békésy (M61), found that it is very important for all scientists to have a few good enemies. When you do your work and write it up, and send it to your friends asking for their comments, they are generally busy people and can afford only a limited amount of time and effort to do this. But if you have a dedicated enemy, he will spend unlimited time, effort, and resources trying to prove that you are wrong, and this will greatly help you to bring out your best performance.[64] Günter Blobel (C99) also thinks that his difficulties in publishing the innovative findings that ultimately brought him the Nobel Prize helped him to better formulate his ideas. Capable opponents are beneficial in crystallizing new ideas and searching for additional experimental evidence. Nonetheless, Max Planck remarked that a new scientific idea does not triumph by convincing its opponents, but rather because its opponents eventually die.[65] The story of Peter Mitchell's (C78) chemiosmotic hypothesis falsifies, in Karl Popper's sense of the meaning, Planck's pessimistic view. The former opponents of Mitchell's ideas eventually became their promoters and championed his Nobel Prize. It did not hurt, either, that one of them was Lars Ernster, an influential member of the Nobel Committee.

To Francis Crick, it was a crucial element of his collaboration with Watson for the double helix that 'If either of us suggested a new idea the other, while taking it seriously, would attempt to demolish it in a candid but nonhostile manner.'[66]

Stubbornness and perseverance, along with an amount of self-confidence and defiance, are needed for even the best scientist to get through with his findings, especially if they are truly new. While Paul Berg (C80)[67] was a PhD student, he worked on a problem that at the time was a central concern in biochemistry. Berg's results were at variance with those of the distinguished biochemist Vincent du Vigneaud (C55). When Berg presented his work at a national meeting du Vigneaud stood up and questioned him aggressively about his data, but Berg defended his conclusions just as aggressively. Later du Vigneaud offered Berg a position, not knowing that he was still a graduate student.

In 1952 Feodor Lynen (M64) and Fritz Lipmann (M53), both future Nobel laureates, published a theory of how to explain a particular biochemical reaction central to a lot of metabolism. Berg found their systems very interesting and decided to work on them. He purified the system more than the original authors had done, showing that their results were an artifact and that the reaction could be explained in a new way. Lipmann and Lynen asked Berg to present the new results at the next biochemistry meeting. After his presentation, Lynen told Berg that it was the first time that he had been proved wrong.

The toughness of some Nobel laureates, their drive and determination, is legendary. Derek Barton (C69) had a reputation for being tough, although he did not think himself as such. He expected his students to work only as long as he did, and not much longer than the professor as in most parts of the world.[68] Donald Cram (C87) was also known to be tough. He thought[69] that being a professor would never be as tough as were his early experiences in other workplaces. Out of his 220 co-workers over the years, he should have fired five or six early, but did not. For one student he tried the following. The student went home for two weeks at Christmas time. Cram worked at the student's bench, on the student's project, recording his results in the student's notebook, and advanced the student's project substantively. Cram hoped to shame his student into working harder by setting a good example. (It was not successful.)

When Herbert Brown's (C79)[70] co-worker tested 57 substances for a certain reaction and 56 of them showed consistent behavior, the co-worker suggested dropping the exception and publishing the results from the other 56. The 56 substances that gave the expected reaction were standard materials in Brown's lab. They had been tested and purified by a careful procedure. The odd 57th compound may have contained impurities and it would have been easy to write it off and go ahead with their usual procedure. However, Brown insisted on repeating the experiment. It was in his nature not to rest until he got to the bottom of things. As it turned out, the 57th compound behaved in an understandable way when they let it react for a more extended period of time than the rest. It contained a double bond for which it took a longer time to add a B–H bond, which it did, and this recognition led to the discovery of a new chemical reaction, hydroboration, and to Brown's Nobel Prize.

Industrial and technological inventors and innovators seldom win Nobel Prizes. Guglielmo Marconi and Carl F. Braun were notable exceptions, winning jointly the physics prize in 1909 for wireless telegraphy; Gustaf Dalén (P12) was another (see also below concerning misguided Nobel Prizes).

In chemistry, industrial and technological innovations have won the prize as seldom as in physics. In 1931, Carl Bosch and Friedrich Bergius were awarded the Nobel Prize for the development of chemical high-pressure methods. Technology was also mentioned in the citation of the 1963 chemistry prize. It was divided between Karl Ziegler of the Max

Planck Institute for Carbon Research in Germany and Giulio Natta of the Milan Institute of Technology 'for their discoveries in the field of the chemistry and technology of high polymers'.

The physics prize for 1939 was awarded to Ernest O. Lawrence of the University of California at Berkeley 'for the invention and development of the cyclotron and for results obtained with it, especially with regard to artificial radioactive elements'. The cyclotron accelerates particles on a spiral path, which causes them to possess high energy and thus enables them to incite nuclear reactions. In 1939 over 20 names appeared among the nominations for the physics prize and Ernest Lawrence was nominated twice. The Nobel Archives store a letter of 3 January 1939,[71] in which M. Stanley Livingston of the Massachusetts Institute of Technology and former doctoral student of Lawrence describes the development of the idea for the cyclotron. This letter was a response to an inquiry from C. J. Davisson (P37) who, apparently, did not find Livingston's report convincing enough reason for nominating Lawrence. There are two interesting features in Livingston's evaluation, and he was well informed since Lawrence had charged him with building the first ever cyclotron in 1930. One is that he traces Lawrence's idea about the cyclotron to an earlier work and the other is that the notion of the importance of a team of scientists comes through in a forceful way even at this early stage of the physics awards.

Lawrence's idea for the cyclotron apparently came from a paper about the production of high energies[72] in a linear accelerator. Leo Szilard had also read this paper and had filed a German patent in 1929 on the cyclotron idea. It was a typical Szilard way of dealing with a problem: have an idea, patent it, then move on. Although Szilard sent a copy of his patent to Lawrence, the latter is not known to have ever commented on it. In any case it was only Lawrence who was brave, determined, and practical enough to accomplish the deed. Livingston took his PhD degree on the principle of the cyclotron and an experimental model in 1931. Then he stayed on at Berkeley to build the first working unit in 1932. Livingston stresses that many people contributed to the successful development of the cyclotron and he suggests mentioning them in any historical discussion of the cyclotron, if not by individual names at least as a group. The invention of the cyclotron could have been awarded a joint Nobel Prize between an individual and a group, providing an early example of the potential benefits of relaxing the three-person rule of the *Statutes*.

Areas

The three science Nobel Prizes cover large areas of science and each can be subdivided into smaller, competing and overlapping areas. There is no indication that, for example, within physics there would be any regular distribution of prizes among, say, solid state physics, elementary particles, radio-astronomy, etc., or in chemistry among organic, inorganic, physical chemistry, and so on. On the other hand, it cannot be excluded that various representatives of the different sub-fields within the Nobel committees as well as in the Science Academy and the Karolinska Institute would not be watching out for proper recognition for their own fields. Many other things being equal, such considerations may enter the decision process. Due to the large number of nominations for the prize in physiology or medicine, they are divided into the following six groups: I. Anatomy, Histology, and Genetics; II. General Biology, Physiology, Physiological Chemistry, and Theory of Drugs; III. Pathology and Pathological Anatomy; IV. Medicine, Surgery, and Therapy; V. Bacteriology, Ethology, and Hygiene; VI. Immunology. Basic science prevails over clinical medicine among the Nobel Prizes in physiology or medicine. Relatively seldom have there been Nobel Prizes given for new drugs. The finding of streptomycin, as 'the first antibiotic effective against tuberculosis' was one of the exceptions. It was discovered by the graduate student Albert Schatz and his mentor, Selman Waksman (M52). Only Waksman received the Nobel Prize for it, though. When James Black, Gertrude Elion, and George Hitchings, the discoverers of quite a few important drugs, received the Nobel Prize in Physiology or Medicine in 1988, no specific drugs were mentioned in the citation. It said, 'for their discoveries of important principles for drug treatment'.

It was not only important principles but also powerful drugs, being used by millions all over the world, that came out of the 1988 laureates' work. They included the beta blocker propranolol hydrochloride (Inderal®) and the antiulcer cimetidine (Tagamet®) by Black, 6-mercaptopurine (Purinethol®) for children with leukemia, the immunosuppressive azathioprine (Imuran®), allopurinol (Zyloprim®) for cancer patients and for treating gout, and the antiviral acyclovir (Zovirax®), by Elion and

Hitchings. The first drug approved for treating AIDS patients, azidothymidine, was also developed by their principles.

Sune Bergström, Bengt Samuelsson, and John Vane were awarded the physiology or medicine prize in 1982 for discoveries of pharmaceutical significance on prostaglandins. There was tremendous societal importance in their work. A most active research area was the fertility field, which became part of a World Health Organization project. Bergström served as chairman of the prostaglandin area, which included the interruption of pregnancy. They had clinical trials in many countries and especially in developing countries. One of their substances was the strongest drug known to stop bleeding after delivery. About 600 000 women died annually from bleeding after giving birth. It was especially striking in India. Bergström remembers an airplane crash in which 20 passengers were killed, and the kind of airplane involved in the accident was grounded for five years. The number of women who died every year from after-delivery bleeding corresponded to one full plane crashing every day. The special program in human reproduction turned out to be very successful. As the program expanded, eventually it was providing about a million women with prostaglandins upon their delivery.

Earlier, the discovery of penicillin, with its impact on infectious disease, was honored in 1945 (Ernst Chain, Alexander Fleming, and Howard Florey), and so were the achievements on the antibacterial protonsil (Gerhard Domagk, M39) and on antihistamines (Daniel Bovet, M57). Curiosity is an important ingredient, even in the discovery of successful drugs and new medical devices.[73] It is a widely held belief that work on penicillin was started as a contribution to the war effort at the beginning of the Second World War. However, work actually began in 1938, motivated by scientific interest, and it was only when its potential as a chemotherapeutic agent for treating infected war wounds was recognized that it became part of the war effort. It is well known that penicillin itself was discovered in 1923 by Alexander Fleming. He found that a mold, contaminating a Petri dish on which staphylococci were growing, dissolved these bacteria. The penicillin mold exuded a substance, which Fleming called penicillin. At that time, however, its potential as a chemotherapeutic agent was not recognized.

It may be interesting to draw up Nobel lineages for various areas but it is doubtful whether meaningful conclusions may be made on their basis, and the overlaps among various fields make such an exercise even more doubtful. The notion that some fields come up at certain intervals may be as valid as the observation that certain sub-fields tend to be recognized by a cluster of Nobel Prizes within a short period of time.

Misguided prizes

Misguided prizes are rare. In 1927, the Nobel Prize in Physiology or Medicine was given to Johannes Fibiger of Copenhagen University 'for his discovery of the Spiroptera carcinoma'. In time it turned out to be a non-discovery, it simply did not exist (see also p. 34). It was a mistake on the part of the discoverer as well as on the part of the awarders of the prize, but it hindered for decades the materialization of other prizes for cancer research. Thus, Peyton Rous (M66), who discovered tumor-inducing viruses in 1910, was not awarded the Nobel Prize until 1966.

Gustaf Dalén's physics prize in 1912 was another benevolent mistake. He received it 'for his invention of automatic regulators for use in conjunction with gas accumulators for illuminating lighthouses and buoys'. That prize was given for sentimental reasons. Dalén had had an accident and had been blinded shortly before the meeting of the Academy. The Nobel Committee for Physics suggested the Dutch Heike Kamerlingh-Onnes for his achievements in low-temperature physics. When the doors were closed and the members of the Academy started the discussion of Kammelingh-Onnes's prize, someone suggested that they defer it until the following year, and give the 1912 physics prize to Dalén instead. And this is what happened. In that year, by pure chance, one of Dalén's colleagues had written a formal nomination for him.

The Nobel Prize to Antonio Egas Moniz (M49) of the University of Lisbon has been controversial. In the mid-1930s he introduced lobotomy, severing the nerve paths to the prefrontal lobes of the brain. The award lent legitimacy to a doubtful surgical procedure used for patients suffering from schizophrenia or manic depression. By the time its inventor died the procedure had also disappeared from practice.[74]

Partners

Scientific partnerships are often reflected in Nobel Prizes, an obvious example being when mentor and pupil share an award. Such was the case when Perutz and Kendrew shared the chemistry prize in 1962 and R. G. F.

Norrish and George Porter shared half of the chemistry prize in 1967. Francis Crick and James Watson, awarded the physiology or medicine prize in 1962, had a different relationship; they were colleagues working together. So were Deisenhofer and Michel (C88), Brown and Goldstein (M85), Bishop and Varmus (M89), and Doherty and Zinkernagel (M96), to name but a few examples.

One of the most beautiful partnerships developed between Gertrude Elion (M88)[75] and George Hitchings (M88) during their 30-year-long joint work at the Burroughs Wellcome Company. When Hitchings, a Harvard PhD who had joined the company in 1942 at 37, hired Elion, with a Master's degree, in 1944, he promised her that there would be no limit to her advancement. This was after Elion had turned down another job offer by a company where she had been told that no advancement would be possible for her because she did not have a PhD degree. Elion was good in synthesis and could read German, which was important in their work. As Hitchings was moving up the company ladder, so was she, taking up more and more responsibilities. Elion did not have her own family although she had a brother with children and they were close to her. Hitchings augmented her background and accompanied her when she went to receive her first honorary doctorate, from George Washington University in 1969. It meant a lot to Elion, who had had to give up her PhD studies back in the 1940s for the lack of support.[75]

The Nobel Prize in Physics for 1915 was awarded to the two Braggs, father and son. This was a unique case, although sons followed their fathers on other occasions, and on one, a daughter followed her parents. This was when Irène Joliot-Curie (C35) and her husband, Frédéric Joliot (C35), received the Nobel Prize for the synthesis of new radioactive elements. Her parents, Marie Curie (P03, C11) and Pierre Curie (P03), shared a prize for their joint research on radiation phenomena (the other half of their prize went to the discoverer of radioactivity, A. H. Becquerel). G. P. Thomson (P37) was J. J. Thomson's (P06) son, Aage Bohr (P75) was Niels Bohr's (P22), and Kai Siegbahn (P81) was Manne Siegbahn's (P24). There was a third married couple that shared the prize, Gerty and Carl Cori (M47), for the discovery of the catalytic conversion of glycogen. In all three cases when husband and wife shared the prize, the announcement was made about so-and-so and his wife, rather than the other way around. Obviously, political correctness was not yet a major issue. There have now been no prizes for married couples for over 50 years.

There are some cases of Nobel intermarriages. This would be less surprising in a small country with relatively frequent Nobel laureates, like Sweden. There, there are Nobel dynasties and, for example, Arrhenius' (Co3) grandson married George Hevesy's (C43) daughter, and in another case one of T. Svedberg's (C26) daughters married the son of a Nobel laureate in literature. Other examples include the following: Henry Dale's (M36) daughter married Alexander Todd (C57); Alan L. Hodgkin (M63) married Marion de Kay Rous, the daughter of Peyton Rous (M66);[76] Frederick Robbins married John Northrop's (C46) daughter and then he himself also became a Nobel laureate (M54); McMillan's (C51) wife was the sister of E. O. Lawrence's (P39) wife;[77] Paul Dirac (P33), already a Nobel laureate, married Eugene Wigner's (P63) sister long before Wigner won the Nobel Prize; and Arthur Schawlow (P81) married Charles Townes' (P64) sister long before either of them was a laureate.

Anticipation

Future laureates have premonitions about their being a candidate. People make guesses and nominators ask the candidates for information, reprints of their publications, and even for suggestions for the best justification to be presented in their nomination. They are seldom above letting the candidate know about their intention to nominate them; the temptation is just too great. The tense wait every October for the news from Stockholm is a dire consequence of somebody's getting close to the Nobel Prize. A famous Harvard professor was known to have come to work in a dark suit the day after the Nobel announcement that did not bring him the prize. (He did finally receive it.) Wolfgang Krätschmer was a viable co-recipient candidate for the fullerene prize and journalists used to camp outside his laboratory in Heidelberg, Germany, every October. He is not the type who expected to receive the prize but could not free himself from the tension around him. So it was a relief when the fullerene prize was given in 1996. Even though he was not among the winners, he could at least get on with his life.

In 1979 the Nobel Prize in Physics was awarded jointly to Sheldon Glashow, Abdus Salam, and Steven Weinberg. Glashow[78] provides a rare glimpse into his suffering from the dread disease of Nobelitis. Based on historical precedents, by 1977 Glashow estimated that he should be among the next in line. He developed insomnia as the next announcement approached. In 1977 the Nobel Prize did come close, at least geographically, when John Van Vleck, a fellow Harvard professor, was among the winners. According to Glashow his Harvard colleague, Steven Weinberg, also developed Nobelitis, which found expression in unpleasant encounters. Remembering past insomnia, Glashow took a sleeping pill for the night of the 1978 announcement. Glashow calls the days of October when the Nobel Prizes are announced 'that dreaded time'. Then in 1979 it was his turn.

Max Perutz (C62)[79] and John Kendrew (C62) heard rumors about their possible Nobel Prize in 1961. They did not want to succumb to a feeling of possible false anticipation, but this became difficult when their secretary brought them two telegrams, one for each. Alas, the telegrams did not come from Stockholm and were about some reprint order. They received the prize the next year.

Donald Cram (C87)[69] measured himself against people he admired. When Vladimir Prelog (C75) received the Nobel Prize, Cram thought he might also have a chance. Cram finds three components important in making a deliberate effort to get a Nobel Prize. The first is to do exceptional research, the second is to bring it to the 'scientific marketplace', which means publishing the results and giving seminars on them all over the world. And thirdly, longevity is very important. Cram said, 'By 1987 when it came, I was not very sanguine about my chances of being so honored.'

Dorothy Crowfoot Hodgkin (C64) was known for her modesty, yet she was aware of her work being worthy of the Nobel Prize.[80] From about 1956 it might have come any year. Hodgkin's wait was the more difficult because several prizes came near, either for their topics or geographically, or both. In 1956, her department head in Oxford, Cyril Hinshelwood (C56), received it; in 1957 it was Alexander Todd (C57) in Cambridge; and in 1958, Frederick Sanger (C58), also in Cambridge, for the chemistry of biological materials. In the late 1950s, Hodgkin's family would congregate around the radio each year to listen to the Nobel announcements and Hodgkin would be disappointed one more time, although her disappointment was confined to the family. In 1962, Perutz and Kendrew won the chemistry prize and Watson, Crick, and Wilkins the physiology or

medicine prize. This came, again, very close. In fact, Hodgkin's pioneering X-ray crystallographic work in biology preceded that of Perutz and
Kendrew. Lawrence Bragg (P15) had proposed her along with Perutz and
Kendrew for the physics prize and Watson, Crick, and Wilkins for the
chemistry prize.[81] In 1964, she finally received the prize and she received
it unshared. Hodgkin was one of very few women laureates in the sciences:

Marie Curie (P03, C11)
Irène Joliot-Curie (C35)
Gerty Cori (M47)
Maria Goeppert Mayer (P63)
Dorothy Crowfoot Hodgkin (C64)
Rosalyn Yalow (M77)
Barbara McClintock (M83)
Rita Levi-Montalcini (M86)
Gertrude Elion (M88)
Christiane Nüsslein-Volhard (M95)

The situation of women in science has been discussed extensively, and one
of the most interesting observations is the conspicuously low percentage
of women in the highest echelons. Even though women's participation in
science has risen considerably, and achieved approximate parity with men
in many areas, it is stagnating at a few per cent among full professors and
members of national academies.[82] The women Nobel laureates, at slightly
more than two per cent, are a scanty share. Rosalyn Yalow (M77)
remarked in her banquet speech in Stockholm:

> We cannot expect in the immediate future that all women who
> seek it will achieve full equality of opportunity. But if women are
> to start moving toward that goal, we must believe in ourselves or no
> one else will believe in us. We must match our aspirations with the
> competence, courage, and determination to succeed, and we must
> feel a personal responsibility to ease the path for those who come
> afterwards. The world cannot afford the loss of the talents of half of
> its people if we are to solve the many problems which beset us.[83]

Longevity is a form of perseverance, and for some Nobel Prizes it was
as much needed as anything else. Peyton Rous (M66) and Karl von Frisch
(M73) were 87 years old when the award finally came. Other laureates
have also noted that they were lucky to have survived long enough for the

Nobel jurors to find their discovery worthy of the prize. In other cases some missing evidence was slow in coming. This is why Paul Boyer (C97)[84] considers himself fortunate that he was still around when John Walker (C97) provided crystallographic evidence for his binding change mechanism and the postulation of rotational catalysis of ATP production. On the other hand, had Boyer disappeared from the scene before Walker determined the structure, it is doubtful that the structure determination alone would have been selected for the award.

Ernst Ruska (P86) received the Nobel Prize a few weeks before his 80th birthday for the first electron microscope, which he had built in Germany in 1933. It took 53 years for the Nobel Committee to award him the prize, and in the meantime electron microscopy had become a real success story. It was a not-too-subtle reference to the time lapse when Ruska started his Nobel lecture by saying that he was reluctant to give the usual scientific 'lecture on something that can be looked up in any modern schoolbook on physics.'[85]

Petr Kapitsa (P78) was 84 when he received the prize for his inventions and discoveries in low-temperature physics. By then he had been away from low-temperature physics for 30 years, as he pointedly declared in the introduction to his Nobel lecture.[86] As early as 1946 P. A. M. Dirac (P33) had nominated[87] Kapitsa and described the essence of his achievements. Kapitsa published these works in the late 1930s, and so his Nobel Prize was more like 40 rather than 30 years overdue. Given these examples, Max Born's (P54) three-decade delay is obviously not a world record. However, the delay becomes conspicuous if considering Werner Heisenberg's (P32) prize, which came only a few years after the 'creation of quantum mechanics', in which Born had had an important role.

At the opposite end is the Nobel Prize for 'the discovery of superconductivity in ceramic materials'. The discovery was made at the beginning of 1986, the report on it appeared later in 1986, and the award was given in the following year to Georg Bednorz (P87) and Alexander Müller (P87), both of IBM Research Zurich. Some thought it was a little too-fast a reaction by the Nobel Committee, although it was very much to the letter of Nobel's *Will*. Nonetheless, it is very rare for the Nobel committees to move so expeditiously. When C. G. Darwin[88] submitted a nomination to the physics committee in January 1940 he stated that he had no doubt that the most important discovery in physics of 1939 was the fission of the uranium atom by Hahn and Meitner. However, he thought it might be a too-hasty award, for it would mean passing over earlier work of value. Hence Darwin nominated someone else for the 1940 prize. Incidentally, Arthur Compton (P27) did nominate Hahn and Meitner for the physics prize for 1940 and so did James Franck (P25).

Joshua Lederberg (M58) finds it reasonable that his Nobel Prize for work on the genetics of bacteria took 12 years to materialize. First it had to be accepted within the inner circles of the scientific community. Then there were the older-line microbiologists, the people who write the textbooks; it took another five years for them to make the corrections regarding sex in bacteria. 'Had the award come sooner that would've meant not only that it was accepted but that it would shoot to the top in priority. There were a lot of other things to give prizes for.'[89]

It is more puzzling that the discovery of the double helix structure of deoxyribonucleic acid (DNA) took nine years, from 1953 to 1962. This may have been caused by the old deep-felt notion that proteins, rather than nucleic acids, were the substances of heredity. This is something that will have to await the release of the relevant material from the Nobel Archives in a few years. Oswald Avery, who first reported that DNA is the substance of heredity,[90] died in 1955 and it might have been somewhat embarrassing to give out a Nobel Prize that he had been denied, so soon after his passing away. That the award was brewing is witnessed by the references to the double helix in Nobel lectures preceding the Nobel Prize to Watson and Crick in 1962. Thus, in 1958 Joshua Lederberg[91] displayed a scheme of Watson and Crick for DNA replication. Then, in 1959, both laureates, Severo Ochoa[92] and Arthur Kornberg[93] presented double helix models in their lectures.

The Nobel Prize to Herbert Hauptman (C85) and Jerome Karle (C85) honored the direct method in X-ray crystallography, more than 30 years after their original work. In this case, though, the method itself would have not meant anything without applications, and they were not quick in coming. Hauptman[94] noted that during 'the first ten years the reaction from the crystallographic community was skepticism at best, hostility at worst'. People started using the method in the mid-1960s, its full significance was not understood until the mid-1970s, and it came to full flower in the 1980s. By the mid-1980s the method was truly established and its use was so widespread that the names of Hauptman and Karle were often omitted in its mentions. Thus the Nobel Prize genuinely surprised Hauptman.

In 1995 Richard Smalley (C96) referred to the anticipation of the Nobel Prize for the discovery of fullerenes this way: 'It makes me, and the people around me, a little nervous every year around the middle of October. This gets to be a little bit less of a problem as the years go by and in time will pass.'[95] As it turned out, the state of anticipation came to a happy conclusion the following year.

4

Discoveries

When we discuss the importance of research projects and discoveries in the context of Nobel awards, we imply that Nobel recognition is given for the most important discoveries. This is probably so, at least in principle, although other considerations also play a role: for example, whether the important discovery can be assigned unambiguously to one or a few persons. In any case, bona fide research is aimed at discoveries rather than prizes. The nature of scientific discovery is an intriguing subject. Johannes Kepler left behind a comprehensive description of all his inner workings in the discovery process. Four hundred years ago he analyzed and described in great detail his thoughts and changes of mood, his exuberance and anguish, during the process of discovering the planetary model. For him the road to the discovery was as interesting as the final revelation. The philosopher A. N. Whitehead, who co-authored *Principia Mathematica* with Bertrand Russell (Literature 1950), said, 'It is more important that an idea be fruitful than that it be correct.' When Aaron Klug (C82) put together his Nobel lecture for publication, the editor wanted to cut out the picture depicting Klug's initial idea of nucleation. He said it was wrong, which it was in its details, but everything essential was in there, so it could show how science is a *process* of establishing the truth. Sometimes you take the right step for the wrong reason.[1]

Style and experience

René Dubos, himself an exceptional researcher at the Rockefeller University, noted that 'scientific creation is a completely personal experience for which no technique of observation has yet been devised. Moreover, out of false modesty, pride, lack of inclination, or psychological insight, very few of the great discoverers have revealed their own mental processes; at the most, they have described methods of work—but rarely their dreams, urges, struggles, and visions.'[2] Little has been done to study the inner processes of making Nobel Prize-level discoveries, or for that matter, of making scientific discoveries in general. Some even think that turning to fiction helps to bring the process of scientific creation to light.[3]

Richard Feynman (P65) recognized the need to tell about the *real* process of discovery. He complained that 'We have a habit in writing articles published in scientific journals to make the work as finished as possible, to cover all the tracks, to not worry about the blind alleys or to describe how you had the wrong idea first, and so on.' He chose the opportunity afforded by his Nobel lecture to narrate the development of the space–time view of quantum electrodynamics, as it really happened.[4]

Peter Medawar (M60) maintained that 'scientists are always dispensable, for, in the long run, others will do what they have been unable to do themselves.'[5] On the other hand, François Jacob (M65) stresses that if somebody else makes a particular discovery, it will not be exactly the same. 'There is style in science too.'[6] Although Francis Crick (M62) believes that, rather than Watson and Crick making the DNA structure, the structure made Watson and Crick,[7] the discovery of the double helix certainly carried their style. It was a master stroke, whereas others might have brought it about in a slower, more stepwise fashion. Emilio Segrè (P59)[8] speculated on how much longer a discovery, if it had not been made by that particular person, would have taken for somebody else to make. Special relativity without Einstein (P21) would probably have been found within a year or two. However, quantum theory might have been delayed by five years or more without Planck (P18). There is general agreement about this; it was uncharted territory and a surprising breakthrough; of all the discoveries, the quantum was the strangest.

Georg Wittig (C79) described research in a rather elusive way when asked about it in connection with his Nobel Prize. He said,

> The path of research rarely leads in straightforward fashion from
> starting point to desired goal . . . chance occurrences along the way
> often enforce a change of course . . . as we come upon various
> points of interest which invite us to linger awhile. Ours, like all
> such rambling tours, possesses that special attraction that comes
> from knowing that the landscape spread out before us will be
> opened to view, not by intention, but by chance and surprise.[9]

Joshua Lederberg (M58)[10] was 33 years old when he received the Nobel Prize for his work on the genetics of bacteria, which he did when he was 21. However, when he attempted an analysis of the process of his own discovery, around the age of 70, he found the task overwhelming.[11]

The finding of the first step in deciphering the genetic code by Marshall Nirenberg (M68) and Heinrich Matthaei raises an interesting question about scientific discovery. Here was a tremendously important question with enormous competition in the search for the answer, and a beginning scientist not only hits on such a seminal problem on his own but also finds the solution.[12] This example will be discussed later in the chapter.

The following anecdote is Gerald Edelman's (M72)[13] example about scientific discovery. Beethoven's landlady says to him, 'Beethoven, get out of my house. Your cat drinks my milk, you throw your laundry in the stairwell, and you pound on the piano all night, I can't sleep.' He says, 'Mrs Schmidt, don't do this to me. You're my inspiration.' And she laughs in response, 'Ha-ha-ha-haaa' (the first notes of the *Fifth Symphony*). That's discovery, according to Edelman, namely, contingency, accident, pattern, preconception, elaboration, and constantly playing back and forth against tradition of some kind; in Beethoven's case, it was Viennese classical music. In tracing a discovery there is an extraordinary complexity and diversity in the history, circumstance, cultural development, and technical skill. Thus, Edelman maintains, it is impossible to lay down any simple rule.

On at least one occasion a previous Nobel Prize generated work leading to a discovery and another Nobel Prize. C. V. Raman (P30) was greatly excited by the news of Arthur Compton's (P27) award for the discovery of the Compton Effect, describing X-ray scattering. Raman exclaimed to his associate: 'Excellent news . . . very nice indeed. But look here Krishnan, if this is true of X-rays, it must be true of light too. There must be an optical analogue to the Compton Effect. We must pursue it . . .'[14]

An example in which the discovery was almost unavoidable concerns the production of poliomyelitis virus in tissue culture,[15] which was a prerequisite for creating vaccines against the terrible disease, polio (earlier called infantile paralysis). John Enders and his two associates, Thomas Weller and Frederick Robbins, of Harvard University received the Nobel Prize in Physiology or Medicine in 1954 'for their discovery of the ability of poliomyelitis viruses to grow in cultures of various types of tissue'. With this discovery they made it possible for Jonas Salk and Albert Sabin to develop their vaccines (see also p. 228). Enders, Weller, and Robbins were working with tissues in culture, employing somewhat different methods from earlier investigators, and had succeeded in keeping the tissues viable for a longer period. They were interested in a few selected diseases, such as infant diarrhea and chickenpox, but not, originally, in polio because so many other laboratories had been working on it that it appeared as a 'bandwagon' topic of the day. It was almost by default that they tried out polio, too. They had some tissue cultures and some poliovirus, so why not put some polio in the cultures? Their success, where others had failed, was quick and complete. In this case the discovery found the researchers, who otherwise were fully prepared for the discovery, having worked out the procedure by painstaking experimentation. Their discovery was also preceded by the most careful review of the literature on viruses and tissue cultures, so there was nothing serendipitous about it. They put the poliovirus into the culture clearly seeing the possibility of succeeding with it.

Recognizing the discovery

A discovery often does happen as a by-product or serendipitous observation. Its importance may not be recognized, especially not at once. There are scientists who had made an important discovery and then moved on with their careers, only to return much later to their early finding. By then they had realized that that early discovery was the most important of their life, and that it might be worth exploring and exploiting it further.[16] When Stephen Berry found the so-called Berry Pseudorotation, he did not even write a full article about it and the discovery was almost buried as a section in a paper about other things.[17] Dan Shechtman unexpectedly discovered the quasicrystals[18] and showed extreme perseverance in having his dogma-breaking observation published,[19] only to then abandon it for years for other, better-funded research. Philip Eaton spent just two weeks making the cubane molecule and is almost irritated by its extraordinary success on the background of his decades of other synthetic organic chemistry.[20]

Harold Kroto (C96)[21] made the first carbon–phosphorus double bond in chemistry, but as soon as it was done he moved on to other areas of research. To the present day he feels that he left too quickly and his contribution did not register sufficiently. From this experience he decided that, were he to make another discovery, he would stay with it for a while. Luckily, it happened, and he decided to continue with the fullerenes for five years after the discovery of C_{60}. Eiji Osawa[22] was the first scientist to come up with the proposal that there might be a highly stable C_{60} molecule of the truncated icosahedral shape. However, he did not find it important enough to write about in English, let alone to follow it up with further work. When Osawa saw the paper in *Nature* about the experimental discovery of the C_{60} molecule,[23] by now called buckminsterfullerene, it was 'the worst day of his life'.[22] When, in the quest for producing the substance, Kroto was pipped at the post by Krätschmer and Huffman,[24] it was Kroto's 'worst day of his life'.[21]

Kary Mullis (C93) recognized at once the importance of his polymerase chain reaction:

> From the very beginning I thought that it would spread all over the world. . . . The same night I thought that if it worked I would get the Nobel Prize, and some day I would walk into the biochemistry department of the University of Zambia, they would know who I am, and they would ask me to say something nice to their graduate students. But I also had this doubt because it was so simple, why hadn't somebody else come up with it before? It took me a lot of wine that night to get to sleep. Then I woke up next day and I still couldn't find anything why it shouldn't work.[25]

At some point in their studies into the fate of chlorofluorocarbons (CFCs) in the upper atmosphere, Sherwood Rowland (C95) and Mario Molina (C95) realized that they were looking at an ozone removal process that was dominant over the processes in the normal stratosphere. That changed the whole exercise, from a scientifically interesting problem into a serious environmental one.[26]

The recognition of the importance of a discovery gives added incentive to further research in the right direction. This is increasingly so when a new research direction is being formed and some recognize it before others. Biochemistry, which is a dominant science area today, developed after Nobel's time. He could have not foreseen it. What is more, even the chemists took a very long time to recognize its importance. Lars Ernster felt very strongly about this issue and diverted criticism that too many chemistry prizes went for biochemical topics. He maintained that

> had not it been for biochemistry, where is the organic chemist who would have conceived the concept of proteins. Has it ever occurred to an organic chemist that there are molecules composed of one thousand amino acids connected with peptide bonds? We learned about these molecules because biochemists had isolated them from nature, and characterized them. There has been a great conceptual contribution by biology to chemistry. We are truly talking about new *chemical* concepts. When we hear this argument that we are giving out too many prizes to biochemists, we always have to remember this.[27]

Chemists used to neglect biological macromolecular substances. They considered them ill-determined and for a long time they ignored proteins and nucleic acids and largely excluded them from the curriculum. However, during the last decades of the twentieth century there was a change in their attitude. Albert Eschenmoser, one of the world's foremost organic chemists, used to provoke Vladimir Prelog (C75), the great natural products chemist, towards the end of his life: 'Vlado, every year during which we did not work on DNA was a wasted year.' Eschenmoser thought that for historical purposes it was important to get an explanation from Prelog, who long resisted giving one. Finally Prelog prepared a one-page statement[28] in which he conceded that they used to consider nucleic acids as dirty mixtures that they should not investigate with their clean techniques.

Paul Berg's (C80) prize-winning discovery was the making of recombinant DNA, whose consequences were enormous. No wonder that Berg is often referred to as 'the father of genetic engineering'. When the discovery was made, Berg recognized its importance and issued a warning in the form of a question, 'Do such experiments cause a potential biohazard for man and his environment?'[29]

The possibility of moving genes to and from all kinds of organisms scared people. They were afraid of creating monsters. The National Academy of Sciences (Washington, DC) asked Berg to form a group of experts; this group issued a moratorium letter and convened the famous Asilomar conference in California in 1975. It brought together scientists from all over the world who wanted to use the technology of genetic manipulation and were concerned about its safety.[30] Guidelines were agreed upon for future experiments. As experience and knowledge accumulated, genetic manipulation was found to be safe. After a slow and cautious start, the work expanded, then exploded.

The important and the possible

A most important problem may not be the most difficult, and one may spend the same amount of effort on solving an unimportant problem as on solving an important one. Lawrence Bragg taught his pupils 'to concentrate on problems of central importance, to approach them directly, to waste no time on trivialities . . .'[31]

The physicist Rutherford's (Co8) dictum was, 'Never attempt a difficult problem,' but it is 'an attribute of genius to see which of the problems are not really difficult.'[32] Recognizing that a problem is not difficult is a necessary but obviously not a sufficient condition for making a good choice.[32]

Derek Barton (C69) paid careful attention to the problems he picked to investigate. He was concerned both about the importance of his problems and that they be solvable. He said that the right problem 'will be significant when you have solved it and will be solvable with the means at your disposal. So it's not good picking too large a problem or a problem where there are no tools to tackle it.'[33]

The question then arises of how to judge a problem to be timely? How to avoid making premature discoveries? Oswald Avery may have determined that DNA was the substance of heredity a little prematurely and William Astbury may have been a little ahead of time in studying fibers. Had he waited a couple of decades he could have had the benefit of additional techniques in his studies. On the other hand, Kroto and Smalley were almost too late in discovering buckminsterfullerene, and when Krätschmer and Huffman produced measurable amounts of C_{60}, they, too, were late for their own sake, though just in time for Kroto and Smalley. Marshall Nirenberg (M68) was exactly on time with his discovery; the world was ready and waiting for the cracking of the genetic code. He was almost a little too early for himself as his racing partners were better prepared for taking up the challenge than he was in the wake of his initial discovery.

In addition to his prize-winning research on vitamin C, Szent-Györgyi (M37) made his most important contribution in the biochemistry of muscle action. He was a great romantic outside the lab but a realist in it.[34] His outside attitude may be best characterized by what he said about fishing: 'Whenever I go fishing I use a big hook so that the fish I don't catch should be a big one.' However, he did not bring this romantic approach into the laboratory. He knew what was important and he knew what was possible, and he made a compromise between the two. Once he told one of his closest associates: 'I enjoy muscle research but my dream would be brain research. Do you know why I don't do it? The reason is that . . . our technical capabilities would not allow me to reach truly important results.'[35] The fisherman Szent-Györgyi earned admiration for his romantic, but alas hopeless, acts during the German occupation of Hungary toward the end of the Second World War. Although he did not accomplish his goals, he showed a different pattern of behavior from the dominating one. In science he earned admiration only for what he did accomplish. Peter Medawar (M60) noted, 'No scientist is admired for failing in the attempt to solve problems that lie beyond his competence.'[36]

Incomplete information

It is essential in scientific research to make decisions on the basis of incomplete information. This is in contradiction to what we usually learn in school where our teachers often tell us to gather all the information before making our decisions. 'Do not jump to conclusions', we hear in our everyday life. Yet in scientific research, especially if it is an excursion into the unknown, not all decisions can be so informed. Of course, the amount of available information that may appear insufficient to some may be viewed as sufficient by others for reaching a decision or for taking the next step.

In his grade school physics class, Melvin Calvin (C61) was in the habit of responding to the teacher's questions almost before the question was out of his mouth. The teacher was very unhappy with him and told him, 'You'll never make a scientist because you don't allow all the information to be presented before you decide on an answer.' Later, Calvin told his own students that

> it's no trick to get the right answer about some scientific question
> when you have got all the data. A computer can do that. A real
> trick is to get the right answer when you've only got half the data
> and half of what you have is wrong, and you don't know which half

is wrong. Then when you get the right answer you're doing
something creative. . . . That philosophy can lead you also into
great troubles, and it frequently does but you can make advances
that way because then you won't be bothered too much by the
dogma of the day.[37]

The discovery of buckminsterfullerene, C_{60},[23] at Rice University in 1985
provided a beautiful example of the masterful utilization of incomplete
information. It also illustrated the difficulties that such incomplete information may generate. The discoverers proposed a three-dimensional
structure in the shape of a truncated icosahedron based on mass spectrometric evidence and utilizing symmetry considerations. Some purists
accused the authors of having overstepped important limits in assigning a
structure on the basis of the available incomplete evidence.

In 1984, an Exxon research group published a mass spectrometric
study of vapor evaporation from graphite.[38] They determined the relative
abundances of the large number of various species present in the vapor as
products of graphite evaporation. In hindsight the C_{60} peak appears conspicuous in their mass spectrum, but was not noticed and discussed by the
original investigators. As it turns out, a young postdoctoral fellow, Robert
Whetten,[39] working for a period at Exxon, had been asked to examine the
mass spectra to see if he noticed anything special. He did not, and nor
could Roald Hoffmann (C81) and Dudley Hershbach (C86) give him any
useful advice when he consulted them about the question. Whetten maintains that it is a taboo to go for a single peak in the mass spectrum and
assign it a structure. If somebody working in mass spectrometry publishes
a structure, saying that this was the structure for this mass, that person
would be excluded from the community and regarded as untrustworthy.
So the feeling was, after publication by Kroto and co-workers of the
Nature paper[23] which proposed the C_{60} structure, that the authors had
violated an important taboo. There were two main differences between
the Rice and the Exxon studies. One was that at Rice they varied the
experimental conditions, and two, at Rice they suggested a structure.
Once a structure was suggested, they coined a name, and the discovery
was complete. There was some risk involved, to be sure, but as it
happened, they proved to be correct in their initiative.

Derek Barton (C69) used the expression 'gap jumping'[40] for con-
necting remote observations. His insights allowed him to see relationships

between facts that escaped others. Albert Szent-Györgyi put it succinctly: 'Research is to see what everybody has seen and think what nobody has thought.'[41]

Paul Boyer (C97) made almost quantitative estimates of the available information, on whose basis he would be willing to take the next step, clearly seeing the risks involved in it. He estimated the degree of risk involved and worked out some analogies to illustrate this. 'If I want to test myself to see how sure I am of something I would say, "Would I bet a granddaughter on it?" I would not bet a granddaughter on the rotational catalysis; I have barely reached the stage to bet all the scientific support in this country on the validity of rotational catalysis, but not a grand-daughter.'[42] The hypothesis of rotational catalysis was the discovery that earned Boyer his share in the 1997 Nobel Prize.

Max Delbrück introduced the principle of 'limited sloppiness'. He said, 'If you are too sloppy, then you never get reproducible results, and then you never can draw any conclusions. But if you are just a little sloppy, then when you see something startling you . . . nail it down.'[43]

George Olah's (C94) discovery of the carbocations was made possible by the superacids that gave those carbocations a longer life, hence allowing them to be observed. In the area of superacids Ronald Gillespie was another pioneer. At one time they were both in Canada and Gillespie had the only nuclear magnetic resonance (NMR) machine around. Olah would send him his materials and his technician to run the NMR spectra to identify the molecules present in Olah's materials. Some of Olah's samples looked like black gunk that Gillespie would have never let his students put into the NMR machine.[44] While Gillespie insisted on purity, Olah discovered the carbocations in those seemingly dirty samples. This was an excellent example of Max Delbrück's 'limited sloppiness' principle.

Overly pedantic work may be counterproductive in noticing new things, and waiting for 100% information may prevent one from making bold, innovative observations and conclusions. Also, if there are really big effects observed, minor inconsistencies may be overlooked at the initial stages, and refined later. Stories about Linus Pauling (C54) and Albert Einstein (P21) show that when they had great ideas, they did not let themselves be deterred by some inconsistencies with experimental observation. They took the risk and eventually proved to be right.

Linus Pauling's discovery of the α-helix[45] is an instructive case in studying the nature of scientific breakthrough. It was not a chance discovery because it was being consciously sought by two groups at the time. It had been known from the early 1930s that polypeptide chains could appear in two versions. Today we call the coiled form α-helix and the extended one β-pleated sheet. In order to ascertain the structure of the coiled form, Pauling first determined the configuration of several amino acids that are the building blocks of a peptide chain. Then he established the planarity of the peptide bond. Finally, he remembered a mathematical theorem that the most general operation that converts an asymmetric object (such as the amino acid units) into an equivalent asymmetric object is a rotation–translation, producing a helix. All this happened over the course of about 15 years. Using all this information he built a model that was in reasonable, though not perfect, agreement with the available X-ray diffraction patterns of the coiled form. He did not care that his helix had a non-integer screw and disregarded a marked discrepancy with experimental evidence concerning the meridional reflection that suggested a repeat at 5.1 angstroms. Later Pauling and, independently, Francis Crick discovered that there was additional coiling of the helices in their packing, causing a change in the meridional reflection as observed in the X-ray diffraction experiments. Crick called this a nice example of symmetry breaking by a weak interaction.[46]

The lesson of Pauling's discovery of the α-helix is complex in that he utilized certain pieces of information and ignored others. It is an important part of talent, perhaps of genius, to know what to take into consideration and what to ignore. The tremendous experience and accumulation of information of structural chemistry certainly added to Pauling's ability to distinguish between the essential and the expendable.

There are dogmas in science that it is considered sacrilege to overstep. Yet it happens that a seminal discovery comes about because someone challenges such a dogma. The discovery of parity violation in elementary particle physics provides a beautiful illustration. Handedness plays an important role in ordinary life and in biology. When we view our left hand in the mirror it appears to be our right hand and vice versa. Yet physicists used to believe that the laws of physics did not distinguish between left and right. This geometrical principle is called space-reflection symmetry and it is expressed as the law of parity conservation. Everybody seemed to believe it except the two young physicists, T. D. Lee (P57) of Columbia University and C. N. (Frank) Yang (P57) of the Institute for Advanced Study in Princeton. They raised the possibility that the weak interactions might be changed by the space-reflection operation. They challenged the experimentalists, and Chien-Shiung Wu[47] and her collaborators did the relevant experiment. A dogma that had been considered almost self-evident was thus uprooted.[48]

Evolving discoveries

John Pople's (C98)[49] prize-winning contribution, the 'development of computational methods in quantum chemistry', can hardly be called discovery; Nobel's other term, 'improvement', fits it better. It was a charted work rather than fortuitous finding. Pople mapped his computational revolution during his postdoctoral work, back in 1952. At that time nothing was really possible in terms of practical computations. Pople's general objective has always been to produce theories and associated computational techniques, which would be extensively applicable and illuminate as many chemical properties as possible. This has proved to be a very successful approach and he was helped a great deal by the huge advances in electronic computation. Pople, who carried out his award-winning work in Pittsburg, considers himself 'very fortunate in being in the right place at the right time, and the emergence of electronic computers made it all possible'. This being in the right place at the right time is a recurring characterization of successful careers. Francis Crick referred to this notion by quoting the painter John Minton, according to whom, 'The important thing is to be there when the picture is painted.'[50] Naturally, it takes more than to just 'happen to be there'. Very often the scientist who is in the right place at the right time has done a lot of moving around before he arrives at the right place. Pople could not complain of a disadvantaged situation, having been a student in Cambridge, and worked under the supervision of an outstanding professor. Until 1964 he held appointments in England, but in the United States there was a better audience and better possibilities at that time for computational science than in Britain. Pople did not just happen to be in the right place, he moved there.

Olah[51] made carbocations detectable. The general significance of his discovery was that, in contrast to the generally accepted concept of a hundred years that carbon cannot bind more than four atoms simultaneously, Olah found that under certain conditions carbon can bond five, six, or even up to eight atoms. This opened up very exciting new perspectives in the chemistry of carbon, which is so central to our terrestrial life. It sounds as if it could have been a sudden revelation, but it was not. It was a long process, an evolving rather than a random discovery. It was Olah's luck that his observations made it possible to resolve a major chemical controversy of the 1960s (p. 196).

The discovery of the α-helix, the rejection of parity in elementary particle physics, Pople's computational revolution in chemistry, and Olah's quest for the carbocations, can all be looked at as charted discoveries. The goals were set at the beginning. Other discoveries grew out of much smaller questions. The realization of the sequencing techniques for proteins and for nucleic acids represented such evolving discoveries.

Frederick Sanger (C58, C80)[52] won his first Nobel Prize for working out the technique of sequencing proteins, and the second for sequencing nucleic acids. It was conceivable that Sanger had set out to solve these two important problems, one after the other. In reality, though, it all went by stages. Sanger had taken his PhD degree on protein metabolism in Cambridge, with Albert Neuberger. In 1943 he took a job with A. C. Chibnall, the new professor of biochemistry in Cambridge. He suggested to Sanger that he look at the end groups of insulin since he was interested in the number of amino acids in proteins. Nothing was known at that time about sequencing.

The choice of insulin was motivated by its availability in a pure form. Chibnall had done a lot of analysis on insulin and it had many free amino groups in it. Chibnall put Sanger to work identifying these groups, and he developed a general process for looking at free amino groups. It was called the DNP (dinitrophenyl) method. The peptide bonds in the chain were broken down by an acid, and the DNP was linked to the amino acid by a stable bond. In this way Sanger could identify the end groups. The discovery of partition chromatography, which Sanger applied to separating the DNP-amino acids, was made possible by the work of A. J. P. Martin (C52) and R. L. M. Synge (C52).

From the determination of the end group the project further developed into a method that would do this for proteins in general. Sanger found that there were two chains in insulin. That was the point when Sanger realized that they could get information about the sequence. With some work they could see sequences about four or five residues long. Those were the first sequences determined in a protein. They could separate the two chains of insulin and succeeded in determining the sequence of one of the two chains, 30 amino acids long. Eventually they were able to put the pieces together and determine the complete sequence of insulin.

Arthur Kornberg (M59)[53] and Har Gobind Khorana (M68)[54] moved gradually toward their results related to DNA. Khorana first synthesized ATP and this led him to the synthesis of the more complex coenzyme A. This in turn led to condensing chains of nucleotides, leading further to the synthesis of a stretch of DNA. Kornberg, similarly, was led step by step to DNA. Both Khorana and Kornberg either did not quite appreciate the importance of DNA at the beginning of their studies, or found it too overwhelming to have it set in front of them from the start as their research goal.

Jean-Marie Lehn's (C87)[55] supramolecular chemistry, along with the contributions of his co-winners, Donald Cram and Charles Pedersen, was another example of an evolving discovery. Supermolecules consist of molecules that are capable of separate existence, which in the super-molecule are linked to each other by relatively weak interactions. Years of painstaking work by many scientists in several laboratories led to this new field.

Crystallographic studies also contain cases of long-lasting hard work. Examples include the elucidation of the structures of globular proteins by Max Perutz (C62) and John Kendrew (C62), the enzyme F1 ATPase by John Walker (C97), and a photosynthetic reaction center by Johann Deisenhofer (C88) and Hartmut Michel (C88). The moment of discovery comes at the very end of such a study and the correlation between struc-ture and function gives it meaning. Thus 'evolving' discovery in this case means not so much that the amount of discovery increases as the work goes along, but rather the work evolves, finally leading to the discovery. When Richard Feynman (P65) was asked whether physicists were getting closer to answering the 'big questions' of physics, he replied: 'You ask, Are we getting anywhere? I'm reminded of a situation when I was asked the same question. I was trying to pick a safe. Somebody asked me, "How are you doing? Are you getting anywhere?" You can't tell until you open it. But you have tried a lot of numbers that you know don't work!'[56]

The great crystallographic structure elucidations should not be considered merely as record-setting events for the sizes of the systems investigated. Solving the structure is the most glamorous part of the work, to be sure. The biochemical background is less showy and often insufficiently stressed in the publications.[1] Although that part of the discovery by itself does not bring outside recognition, it is important for science. Aaron Klug (C82) uses this metaphor: 'Research is not just going from mountain top to mountain top, you also have to work in the valleys, and that takes time and freedom.'[1] Scientific achievements are often compared with conquering the highest mountain tops.

Dorothy Hodgkin (C64), on a visit once to North-Bengal University in Siliguiri, India, close to the foothills of the Himalayas, wanted to view some of the high peaks. S. Ramaseshan was in her company and remembered Chandrasekhar's (P83) words:

> The pursuit of science has often been compared to the scaling of mountains, high and not so high. But who amongst us can hope, even in imagination, to scale the Everest and reach its summit when the sky is blue and the air is still, and in the stillness of the air survey the entire Himalayan range in the dazzling white of the snow stretching to infinity? None of us can hope for a comparable vision of nature and the universe around us. But there is nothing mean or lowly in standing in the valley below and awaiting the sun to rise over Kanchanjunga.[57]

Changing paradigms

X-ray determinations give unambiguous structures and an X-ray structure meant more than the mere connectivity order; it is a truly three-dimensional structure. Hodgkin's demonstration of such structures for complex organic molecules changed many of the aspirations of organic synthetic chemists. They had been freed of a great burden of proof for structures, and this constituted a true paradigm change.[58] Before the magnificent advances of X-ray crystallography, chemists argued about the structure of natural products. There was a lot of uncertainty in it, and laborious synthetic work, often for years by whole teams, led to the correct solution. John Cornforth (C75) likened the logic of underpinning organic structures to 'the roots of a large tree: tortuous, tangled, and very strong'. Back in the 1930s, when Cornforth studied the mechanism of organic reactions in those pre-X-ray times, there were already lassos and arrows and dotted lines in usage to mark the changes in atomic connectivity in chemical reactions. However, in assigning these little lassos and lines to atoms, the printers' convenience mattered more than the real mechanism of the reaction.[59] Cornforth is credited with having said, in a heated debate over some important details of the structure of penicillin, 'If penicillin turns out to have the β-lactam structure I shall give up chemistry and grow mushrooms.'[60] He disputes this story but finds it undignified to publicly deny it.[61]

Nirenberg (M68) and Matthaei's breakthrough in deciphering the genetic code was a serendipitous discovery. Nirenberg first made the announcement of his discovery of the 'first word to be identified in the genetic code' to the Fifth International Congress of Biochemistry in Moscow in 1961. To him, in the late 1950s, protein synthesis was unambiguously the hottest field in biochemistry. The best biochemists in the world were working on the biosynthesis of proteins. They had just discovered transfer RNA (ribonucleic acid), and the amino acid-activating enzymes that catalyzed the activation of transfer RNA to link an amino acid to a particular species of transfer RNA. They also knew that proteins were synthesized on ribosome particles in the cells. But nobody knew anything about the messenger. This was the first problem that Nirenberg worked on as an independent investigator at the National Institutes of Health, where he stayed after his postdoctoral fellowship there. He asked himself, 'What chance do I have as a single person against the best people with big groups in the best laboratories of the world who were working on protein synthesis?'[62]

By the spring of 1961 Nirenberg realized that he had a terrific thing to tell to the meeting in Moscow. However, being unknown in the field, in fact in any field, he was scheduled to give a 10-minute talk in a tiny room with a giant-size projector, and there was only a handful of people. Word soon reached Francis Crick (M62), who was chairing a large symposium on nucleic acids, and Crick invited Nirenberg to give the talk again in that forum. When Nirenberg did so, he was overwhelmed by the response. This was indeed the first time that anybody had shown definitively, in an *in vitro* system, that RNA directs protein synthesis. It was obvious that they had the first codon. They had shown that polyuridylic acid, poly U, directs the synthesis of polyphenylalanine, a protein. A series of U's in RNA corresponded to the amino acid phenylalanine. This was the beginning of the deciphering of the genetic code, to determine the translation between the structure of nucleic acids and the structure of proteins. When they added a synthetic RNA, containing only one kind of base of a possible four, to their cell-free protein synthesizing system, it directed the synthesis of a protein consisting of only one kind of amino acid, out of the 20 amino acids. This is what they proved and this is what Nirenberg presented at the meeting.

The 1989 Nobel Prize in Chemistry was awarded to Sidney Altman of Yale University and Thomas R. Cech of the University of Colorado 'for their discovery of catalytic properties of RNA'. This was a paradigm change in that nucleic acids, that is, not only proteins, can also be catalysts in biological processes. This was yet another discovery that became important because of a strongly held dogma of previous times. Some called it one of the two most important discoveries in biology for the past half a century, the other being the double helix of DNA. However, this judgement originated from the previous assumption that enzymes are proteins. Taken against this background, it was a revolutionary contribution. On the other hand, Altman sees it in a more realistic perspective:

> If you look for a proper definition of a catalyst (or an enzyme),
> you won't see its chemical nature defined in any way. A catalyst is
> something that accelerates a reaction but it is not defined as RNA,
> DNA, protein, or something else. Any large molecule, or even
> a small one with the right kind of properties, can be a catalyst.
> From a chemist's point of view our observation was not something
> fundamentally new, but it had many important implications for
> biochemists or biologists. One of my senior colleagues here told me,
> half-jokingly, 'Whatever it is, chemistry or not, it's still great.'[63]

What gave special importance to the realization about the capability of RNA to act as a catalyst was the implication for the question about the origin of life. If RNA can be a catalyst, then the whole life process could be started without proteins. The next step in this thinking was that even DNA was not needed to get life started because RNA encodes the information in the same way as DNA and, if it is a catalyst, it can do everything. Walter Gilbert (C80) named this scenario 'The RNA World'.[63]

Joyous moments

Discovery is a unique experience and so overwhelming that the scientist often feels an instant urge to share the moment. Of course, the break-through does not always come as such a momentous point in time but some-times it is possible to pinpoint it. One of the most remarkable moments in science history was when Louis Pasteur first observed two kinds of small crystals of the same substance that were mirror images of each other. He rushed out of his laboratory into the hall, embraced the first person he met, and exclaimed: 'I have just made a great discovery. . . . I am so happy that I am shaking all over and am unable to set my eyes again to the polari-meter!'[64] And when Pasteur showed the old Jean Baptiste Biot, the dis-coverer of optical rotation, his experiment, Biot said, 'My dear child, I have loved science so much throughout my life that this makes my heart throb.'[65]

Kary Mullis made his polymerase chain reaction work for the first time on the night of 16 December 1983 at the Cetus company and he re-membered it as:

> I was so happy, and there was nobody else in the lab. Only he
> [Al Halluein, the patent attorney] was around, and I had to tell
> someone that it worked, and it was he. Al was a southerner and he
> was a friend of mine and he recognized at once that it was going to
> be the most interesting thing he has ever patented.[25]

In the course of the determination of the structure of the photosynthetic reaction center, Johann Deisenhofer (C88) was building models on his computer screen. As the model was emerging he experienced the most exciting moments of the whole project. The culmination came when he noticed an overall symmetry in the huge system. At that point he was sitting in a dark room, interacting with the machine. His first reaction at the moment of discovery was to light up a cigarette, although he had quit smoking some time before. He just could not live with the excitement without doing something like that. It proved to be a bad move because it was very difficult for him to stop smoking again. His second action was to call his co–investigator, Hartmut Michel (C88), and he showed him the whole thing. Deisenhofer says,

> It was very nice to have a colleague like him who in many ways
> complemented my expertise. We could give each other many things

in the course of this work. It was a relationship of complete trust. When such a story becomes known, there is always a temptation to claim the whole fame. This did not happen between us and I'm very glad that it didn't because it could've ruined everything.[66]

During his muscle research Szent-Györgyi made a new discovery, that a fiber drawn from a complex of actin and myosin showed a contraction in the presence of ATP, just as a muscle fiber showed such a contraction. He was already a Nobel laureate yet he retained the enthusiasm of a beginner in his newly found field: 'To see them contract for the first time, and to have reproduced in vitro one of the oldest signs of life, motion, was perhaps the most thrilling moment of my life.'[67]

When Kroto (C96) and his colleagues had made the HC_7N molecule, and had recorded its spectra, he went to the Canadian observatory in Algonquin Park in 1977 to try to detect it in interstellar space. Years later, even after the discovery of buckminsterfullerene, he felt he had the most exciting and cathartic moment in his life when HC_7N came on the screen from space.[21]

The taciturn Paul Dirac (P33) was an exception. Once Richard Feynman (P65) asked him how he felt when he discovered the Dirac equation. Dirac's answer was, 'Good.' End of conversation.[68] The double physics Nobel laureate John Bardeen (P56, P72) was a quiet and modest man. The day they discovered the transistor at Bell Labs, he went home in the evening and told his wife, 'We discovered something today.' Upon the second prize-winning discovery a colleague remembers meeting Bardeen in the hallway of the physics building of the University of Illinois. The colleague sensed that Bardeen had something to say, but it took some time before he spoke up: 'Well, I think we've explained superconductivity.'[69]

Sanger ascribes success in science to being interested in the work. 'Do what interests you. Most of the satisfaction is from the fun of exploring, doing things that nobody else has ever done before. That to me is much more exciting than winning the awards, though that is very nice too, and it helps one in one's career.'[52] Here, then, is Carleton Gajdusek's (M76) *ars poetica*: 'I've always played in science. I've never worked. I don't treat science seriously, I think it's a joke.'[70] Emilio Segrè (P59) compared the trip to Stockholm with the discovery: 'It's nice to receive a prize but to make a discovery is very, very thrilling.'[8]

For Marshall Nirenberg (M68),[62] deciphering the genetic code was a tremendous experience. He asked himself the question of whether the code was the same in bacteria as in amphibians and in mammals? They prepared transfer RNA, they did all the experiments, and they found that the code was the same. It is essentially a universal code. Nirenberg was familiar with, and understood, evolution and Darwin but this was on a different level. He found that looking out the window, seeing the trees and seeing the squirrels, and knowing that the genetic code of these organisms was the same, or essentially the same, as the genetic code in him, was a very powerful philosophical concept. This realization of the unity of nature had a profound effect on him, one that has not waned during his entire career.

5
Overcoming adversity

In American academia, an ideal career starts with having an academic or professional family, being educated in a fine liberal arts college, doing graduate studies at Harvard, Stanford, Princeton, or suchlike, and then doing postdoctoral work at a similar institution. This helps to build up a network and self-confidence. Then there is a job at a top research university with a large start-up package, followed by major grants. In other countries there are corresponding paths, but this kind of background is not typical among the sampled Nobel laureates. Most of those I met had become laureates during the last three decades of the twentieth century. They come from diverse backgrounds and many have struggled with various handicaps. This hardship came in many different ways and at different stages of their lives. For some it was in their childhood, for others, it was in mid-career. Few of them were not exposed to some kind of severe challenge.

Early hardship

Many of the Nobel laureates spent their childhood in the 1930s and 1940s. This meant economic hardship during the Depression, the persecution of Jews in Germany and elsewhere in Europe between the two world wars and through the Second World War, and the struggles with anti-Jewish discrimination through the early 1940s in the United States. The Second World War interrupted early scientific careers even if it was not active persecution. Loss of jobs for parents, loss of father, and poverty were frequent. Another characteristic is migration, which is often a scientist's fate, but many were forced to do it. In the following there will be no biographies, merely the highlighting of a few striking features. It is impossible to draw any conclusion as to whether a given laureate would have performed in a similar way under vastly different circumstances. Nor will we ever know whether hard conditions or persecution—let alone annihilation

—prevented other children and youths from a life path of outstanding scientific performance. In fact, the question is not really 'whether', but 'how many?'

Leon Lederman (P88) read a story from Aldous Huxley, called 'Young Archimedes', which has stayed with him for his entire life. In his words,

> An English mathematics professor is vacationing in some remote
> part of Italy. On a walk he sees a farm child near the river with a
> chopstick making triangles. He looks at the triangles and he realizes
> this child is about to prove Phythagoras's theorem, $a^2 = b^2 + c^2$.
> He starts to talk with the child. The child is at first shy. Eventually,
> over the long summer, he teaches the child formal mathematics,
> geometry, algebra, trigonometry, and calculus. The child is a
> genius. There is nothing too fast for him. At the end of the summer
> the professor goes to the farmer and says I would like to take your
> child to England, to educate him. He will have the best food, the
> best schools, he will come home four or five times a year for
> holidays; I think he will be a great man. But the farmer says, I
> need him to help me in the farm work. The end of the story is
> the English professor and his wife go away and the child waves.[1]

Lederman asks the question, 'How many Amazon kids, how many African kids, how many children in some remote village in China and India get lost and could be a Newton or an Einstein?'

I do not believe that handicap is a prerequisite for the success of Nobel laureates, rather, they prevailed in spite of these handicaps. We like to think that future generations may never experience the hardships that will be mentioned below. However, this is a biased wishful thinking. Much of the world today lives under conditions that may be as testing for some future Nobel laureates as those experienced by the ones who made it for our discussion.

It is also possible, though, that these hardships provided an important component in building character and developing drive that is beneficial for doing science. Donald Cram (C87) had strong views on this:

> my early years taught me how to handle adversity and stress,
> how to respond positively to challenge, to be self-reliant, self-
> determining, and individualistic. As importantly, my early years
> fostered enterprise and creativity and closely linked hard work to
> reward. I believe today's youth compared to my generation are

deprived in one very important respect—they are not challenged, tested and graded enough; life has been too easy for them, they too often have to learn to handle adversity too late in life; they understand too much the 'carrot' but not enough the 'stick' side of incentive. The distillate of my remarks here is that my early environment stimulated self-discipline, whereas the current environment is too rich in self-indulgence.[2]

I have come across laureates who originated from backgrounds that would qualify as ideal conditions. Alfred Gilman (M94)[3] grew up in a comfortable academic family. His father, also named Alfred, was a pharmacology professor at Yale University, a rather exceptional Jewish career in the 1930s. He and his colleague Louis Goodman co-authored a major textbook[4] and Goodman became his son's middle name. According to Michael Brown (M85),[5] Gilman is the only person who was named after a textbook. Another example is Kenneth Wilson (P82), whose Harvard professor father, E. Bright Wilson, will be mentioned among the mentors to Nobel laureates. Carleton Gajdusek (M76)[6] grew up in a highly intellectual and flamboyant environment in which he became acquainted with great scientists and their work early on. John Polanyi's (C87) father, the Manchester professor, Michael Polanyi, will also turn up among the Nobel laureate mentors in Chapter 8. However, the Polanyi family had less than ideal conditions; when John Polanyi was about five years old they had to flee Nazi Germany.

For most Nobel laureates there were tests of will power and perseverance at some point or another in their path. The Depression of the 1930s remained a bitter lifetime experience for many, including Bruce Merrifield (C84)[7] and Daniel Nathans (M78).[8] Herbert Brown's (C79)[9] parents were born in the Ukraine. When they emigrated to the United States his five-year-old sister had a virus infection of the eyes, and was refused admission. Thus they went to London, where Brown's father became a cabinetmaker. Brown was born in London in 1912, but when he was two years old they finally moved to the United States. They lived in poverty and his schooling was interrupted for economic reasons. At the University of Chicago, Brown was in a tenure-track position, but then found out that he had no hope of obtaining tenure there. This was in 1943 and he went to Wayne University as an assistant professor. Neil Gordon[10] was in charge of the chemistry department. They had no PhD program and Gordon wanted Brown to help develop one. The teaching load was 18 hours, but Brown was promised 12 hours to give him a chance to do research. Wayne operated an evening school and Brown arranged to teach from 6 p.m. to 10 p.m. in the evening, three days a week. When his colleagues also wanted a reduced teaching load, they introduced a sabbatical year in residence, every three years on a rotating basis, when they would have only 12 hours of teaching. Further, for every paper they published in a recognized journal, the teaching load would be reduced by two hours. Brown published six papers in the first year, so his teaching load was six hours the following year.

Gertrude Elion's (M88)[11] father came to the United States as a small boy from Lithuania and graduated from New York University School of Dentistry in 1914. Her mother came from Russia in 1911 at the age of 14. They were doing well until they lost everything in the stock market at the very beginning of the Depression. Nonetheless, they insisted on their children getting a college education. Elion advanced as far as a Master's degree but had to give up her studies for her PhD because of lack of funds.

Donald Cram, Sune Bergström (M82), and Elias Corey (C90) were among those who lost their fathers very early. Bergström's[12] mother was left alone with three children. Her life was a struggle and she never remarried. Corey was born in the United States; his grandparents had emigrated there from Lebanon, where they had learned to cope with adversity as Christians in a country that was part of the Ottoman Empire. Cram got his education as much on the streets as in any school. For many years he was trying to imagine how his life would have been had he not lost his father and wanted to build up his father's image. By the time he had succeeded, in his mid-thirties, he realized that he had built up an image of himself, a composite of various characters he admired in his readings, and what he had learned from people he did not want to resemble was even more important. Looking back, growing up fatherless meant to him the advantage of testing himself against circumstances, and growing in confidence, skill, and judgement with each encounter with others. His mother was raised in a strict Mennonite faith, which she rejected and escaped from, by marriage, at an early age. She introduced her children to literature and music early on. She would read to Cram the first parts of books, but as soon as they became exciting, she would stop. To find out what happened in the stories, Cram had to learn to read very early. His mother arranged to barter his labor of mowing lawns and emptying ashes for everything from food to dental care to music lessons. She kept the family together until Donald, the smallest child, was 16, after which the family dissolved and he became self-supporting.

Derek Barton's (C69)[13] father was a carpenter who had sent his son to a good school, but Barton had to interrupt his schooling at the age of 16 when his father died. He spent two years in the wood industry and taught himself a lot by reading. George Porter (C67)[14] was brought up in Yorkshire, in the north of England, a poor world at the time of the Depression. He went to an ordinary tin-hut school and some of his friends there were going around without shoes and socks. His school was not at a high level academically and nobody from it had ever gone to Oxford or Cambridge. Most of his friends left school to work and support their families. Porter went to Leeds University at the age of 17. James Black (M88)[15] grew up in the culture of the coalfields in east Fife in Scotland. There were five boys in his family and he was the second from the bottom, so he learned his place in life. His father was a mining engineer who had begun working as a coal miner when he was 12 and used night classes to advance his career. Black took his degree in medicine in 1946. In 1947 he married, and in order to pay off his student debts he accepted an appointment at the Medical School in Singapore, then a British colony. He returned to Britain in 1950. John Walker (C97)[16] was also born in Yorkshire. When his grandfather died he left debts, which meant that they lost their entire business. So Walker was brought up in poor circumstances. The family efforts concentrated on him as the male child and less on his two younger sisters. Nowadays Walker is concerned that gifted children from the poorer north of England should reach Cambridge and Oxford.

Childhood illness was a determining factor in the life of some. Nikolai Semenov (C56)[17] grew up in the Volga region of Russia. He contracted typhus and, unable to attend school for quite a while, turned to books. Marshall Nirenberg (M68)[18] had rheumatic fever in his childhood. In those days nobody knew what caused it and bed-rest was all they could prescribe for him. Rheumatic fever was a big killer of children at the time he had it, for about five years from the age of eight. At one time he spent a whole year in bed. They thought that a warm climate would be better for him, so his father gave up his business in New York and they moved to Orlando, Florida. John Cornforth (C75)[19] entered Sydney University in Australia at 16, and due to a health condition became progressively deafer. He could not use the hearing aids that were available then because the sound was distorted, and he did not lip-read, although he does now. Even if he had been an expert lip-reader, he could not have used it for lectures: lip-reading is a guessing game and not good for learning new ideas. He tells this to deaf children who sometimes ask for his advice about how to get through university.

Henry Taube's (C83)[20] parents were born in Russia but, atypically for an American Nobel laureate with roots in Russia, they were non-Jewish. Their forebears came to Russia from Germany during Empress Catherine's reign. Taube's parents were peasants without any education, except for the reading they were taught for their confirmation, and lived a miserable life under the czars. Taube finds it important to remember this when we consider the history of the Russian revolution. His parents escaped from Russia in 1911 and went to Canada, settling in Winnipeg, Manitoba, then in Neudorf, Saskatchewan. His father worked as an unskilled laborer, then as a farmhand, and his mother cleaned houses. Henry was born in a rented sod hut. When they had accumulated some means, his father rented a farm and they lived in a two-room shack in nearby Grenfell, which was their home until Taube was 13, at which point he left for Lutheran College.

Vladimir Prelog (C75)[21] grew up in Sarajevo, Bosnia-Herzegovina. His father was a Croat from Croatia. The Croats were Catholic, and as a child Prelog never played with a Moslem child, never played with an Orthodox (Serbian) child, and did not even play with Croatian children whose families were originally Bosnian. Prelog studied in Prague, taught in Zagreb, and in 1941 moved to Zurich and stayed there for the rest of his life. Prelog's parents married very young and divorced early. He was tormented by the fact that his mother did not contact him between the ages of 10 and 27.

Roald Hoffmann's (C81)[22] schooling was another casualty of war. He avoided Auschwitz by hiding with his mother in southeastern Poland. The Germans killed his father, and for years his schooling was chaotic. First, there were a few months in a Ukrainian school in Złoczow. Then the second and third classes were in a Catholic school in Krakow, in

Polish. His fourth class was taught in Yiddish in a displaced persons' refugee camp in Austria. Then he was taught a little in German in Germany, and in the fifth and sixth grades everything was taught in Hebrew in Munich. He was a teenager when the family emigrated to America, and he did not own a book until he was 16 years old.

Career hurdles

The Second World War found future Nobel laureates on both sides. George Porter (C67) served in the British Navy and Manfred Eigen (C67) in a German auxiliary anti-aircraft unit, the 'Luftwaffehelfer', for the last three years of the war. He once thought of becoming a musician, but as he could not play during the war years, Eigen used his free service time for studying. Some fellow enlisted men ridiculed him, and his superior officer was especially upset when he noticed that Eigen was learning English. One war before, another German, Georg Wittig (C79), had had a similar fate. He was the son of a fine arts professor and might have chosen a similar career. He became an accomplished pianist and excelled in music, but he also loved chemistry and decided to study it at university. During his undergraduate studies in the First World War he was drafted into the German army; capture followed and he spent the remaining time in British captivity.[23] After the war he continued his studies in chemistry, but music was no longer a viable alternative. Günter Blobel (M99),[24] whose father was a veterinarian, spent his first years in Silesia, which is Poland today but was then part of Nazi Germany. The family left Silesia at the end of January 1945, driving away in their car as the Russians approached. They did not go far enough because they eventually found themselves in East Germany. Blobel was not allowed to study in a university because his family was considered to be capitalist. He left for West Germany before the infamous wall was erected.

James Chadwick (P35)[25] was born in the Manchester region of England, the eldest son of a poor cotton-spinner and his wife. His parents left him in the care of his grandparents when he was very small, and young James became shy, aloof, and taciturn for his entire life. A bright spot in his youth was being awarded an 1851 Exhibition Scholarship, which enabled him to go to Germany where he worked with Hans Geiger in the Physical–Technical Institute in Berlin. Chadwick was trapped in Berlin when the First World War broke out and spent the four years of the war interned in Germany. He lectured and carried out rudimentary experiments during the internment, and returned to Manchester to work with Rutherford in 1918.

François Jacob (M65)[26] attended medical school in Paris with the intention of becoming a surgeon. The Second World War interrupted his studies and Jacob was among the few men who joined the Free French Forces and General de Gaulle in London. He served as a medical officer in Africa, participated in the Normandy invasion, and was severely wounded. His injuries crushed his hopes of becoming a surgeon. When the war ended and he had recovered from his injuries, he completed his medical studies; his former fellow students were years ahead of him. In 1950 he joined the Pasteur Institute, but to engage in scientific research he first had to return to his studies and obtain a science degree.

Odd Hassel (C69) was working in German-occupied Norway in the early 1940s on his most important project and discovered the concept of conformational equilibrium. At one point the Norwegian Nazis arrested Hassel for his participation in the Resistance movement and turned him over to the German Gestapo. The people at the University of Oslo contemplated a strike, but Hassel sent word from the prison to his colleagues: 'The lectures must go on.'

Rita Levi-Montalcini (M86)[27] grew up in a loving family but her father did not consider science a proper career for women and it took her some effort to convince him that she should study at the university. A great challenge to Levi-Montalcini's determination came with the growing impact of Fascism in Italy. Jewish scientists were losing their jobs, yet Levi-Montalcini did not want to leave the country. Anti-Semitism and ruthless persecution were alien to most Italians, and the predicament of Jews in Italy was different to those in countries like Germany and Hungary. Levi-Montalcini went to medical school but was not allowed to become a medical doctor and started doing scientific experiments at home. This is how she remembers that period:

> Many years later, I often asked myself how we could have
> dedicated ourselves with such enthusiasm to solving this small
> neuroembryological problem while German armies were advancing
> throughout Europe, spreading destruction and death wherever they
> went and threatening the very survival of Western civilization. The

answer lies in the desperate and partially unconscious desire of human beings to ignore what is happening in situations where full awareness might lead to self-destruction.[28]

Hans Krebs (M53)[29] and Gerhard Herzberg (C71)[30] are two examples of scientists who had to leave Germany during the Nazi era because of its Jewish laws. Krebs lost his job at the University of Freiburg and went to Cambridge. Herzberg was not Jewish but his wife was, and by 1935 he had lost his university professorship in Darmstadt for refusing to divorce his wife. According to the Nazi laws a German could not have a university teaching position if he was married to a Jew. Herzberg resigned his position and they left for Canada. Walter Kohn (C98) was born in Vienna. His family was Jewish and after the Anschluss he and his sister fled to England but their parents perished in a concentration camp.

Compared with the Holocaust, the anti-Semitism of the 1930s in America was mild, but it existed to an extent that many today find hard to believe, and many of the Nobel laureates of the second half of the twentieth century experienced it. A few examples follow. Herbert Hauptman (C85)[31] grew up in New York City. If it had not been for City College, Hauptman would not have gone to college, although his parents were supportive and wanted their three sons to go as far as they could. When Hauptman was looking for a job, he experienced discrimination against Jews. In 1940 he learned about a position as a mathematician in the Department of the Navy, and applied for it. He was granted an interview by naval officers, and asked to wait outside. Whether it was done deliberately or not he does not know, but he overheard their discussion referring to him as 'smarty Jew'. He did not get the job. Later, when in the navy during the Second World War, he again experienced blatant anti-Semitism.

Robert Furchgott (M98)[32] took his BSc degree in chemistry from the University of North Carolina at Chapel Hill in 1937. He wanted to do graduate work in physical organic chemistry and had applied to many places, but nothing was coming through. He was top of his class but he was also Jewish. His non-Jewish friend, who was second to him, won an assistantship right away. Finally, Furchgott landed an assistantship in physiological chemistry at Northwestern University through the personal contacts of the Chapel Hill department head. Furchgott's doctoral work was on red cells and proteins, and he wanted to continue in protein chemistry. He applied for a job in the Textile Institute in Washington. Although they would have liked to hire him, the opinion prevailed that there were already too many Jews in the institute. They told him so, and he was turned down.

Arthur Kornberg (M59)[33] did well at New York's City College, but was discouraged from pursuing a career in chemistry. There were no jobs available in chemistry and he would not be employed in any chemical company because he was Jewish. Kornberg was also interested in biology, from which he could have been just as discouraged. Medical schools were largely closed to Jews, and even Columbia University, which was down the street from City College, had not filled a scholarship available for one City College student for the previous nine years, and not for the lack of good applicants. Kornberg went to the University of Rochester, which had a small quota for Jewish students. He was one of two in his medical school class. Kornberg relates a bitter experience of six decades ago as if it happened yesterday. The dean of the medical school at Rochester, George Whipple (M34), bestowed a fellowship year in pathology upon two students, but Kornberg was not offered that fellowship, even though he was the top student in his class. It did not ease his pain when he learned later that the dean discriminated against Italians as well.

Jerome Karle's (C85)[34] parents came to the United States from Eastern Europe. Relatives adopted his mother because his widowed grandfather could not cope with the whole family. Karle went to the Abraham Lincoln High School in Brooklyn and then to City College, from which he graduated in 1937. He hoped to become a medical doctor and applied for Harvard after he had earned a Master's degree in biology there. The dean of the graduate school of Harvard told him, 'We have enough Jews in Massachusetts and I am not going to add one from New York.' Karle then went to Michigan where he became a physical chemist. He wanted to do his graduate work with Lawrence Brockway, a former Pauling student, but although he finished at the top of his class, they did not give him a teaching assistantship. When Brockway protested to the dean, who was a German and had studied under Moses Gomberg, he told Brockway that he never gave teaching assistantships to 'Jews, Negroes, Italians, and women'. Whatever Brockway may have told the dean, Karle had his teaching assistantship by the next morning. This was in 1940.

City College for boys and Hunter College for girls provided tuition-free college education in New York City. The parents of Rosalyn Yalow, née Sussman (M77),[35] also came from poor Eastern European immigrant families. Her mother completed the sixth grade and her father only the fourth grade, but they were determined for their children to have a college education. Rosalyn went to school in the Bronx. They had good teachers and the predominantly poor and Jewish pupils were very motivated. She then went to Hunter College. It was difficult for her to find a graduate school, but eventually she was offered a teaching assistantship at the University of Illinois at Urbana. Essentially it was the war that made it possible for her and many other young Jewish students, men and women, to enter graduate school. Both Paul Berg (C80)[36] and Leon Lederman's (P88)[37] parents were Jewish immigrants from Russia. They both took advantage of the tuition-free college education in New York City. Harold Varmus (M89)[38] and Stanley Cohen (M86)[39] are further typical American Nobel laureates of East European Jewish ancestry. After the Second World War anti-Semitism in America gradually diminished, and even during the war the situation for both Jews and women eased due to labor shortages.

César Milstein's (M84)[40] father went to Argentina when he was 14 years old, from a little village in the Ukraine. Milstein's parents were supportive: his mother typed his Argentine PhD thesis, and his father offered him economic assistance so that he could dedicate himself full time to his research. This Milstein refused because he wanted to be independent even though in those days there were no scholarships in Argentina for research students. He and his new wife had to work to support themselves. After taking his PhD from the University of Cambridge in 1961, he went back to Argentina and stayed for about two years. At first he was happy there; they did good work and published good papers. However, after the military coup conditions deteriorated, and the director of the institute where Milstein worked was persecuted. Milstein protested. He was head of a division and four members of his staff were dismissed. It was clear to him that the authorities wanted to get rid of him; for them all the scientists were being nuisances by protesting against the persecution of the director. Someone called Milstein, obviously a Jew, must be a communist. In fact, they were all considered to be communists. Milstein left Argentina and returned to Cambridge.

George Olah (C94)[41] had embarked on a successful career in academia and had a well-established position in Hungary. Following the crushing of the revolution in 1956, he and his family emigrated to North America, where Olah had to rebuild his career. They first went to Canada, then to the United States, and he spent eight years at Dow Chemical before he could rejoin academia.

Daniel Chee Tsui (P98)[42] was born in Henan, China. He left his village and family in 1951 and went to Hong Kong, where he graduated from high school in 1957 and continued his studies in the United States.

Ahmed Zewail (C99)[43] spent his first 22 years in Egypt. He received a traditional education, which was good but irrelevant for his career. He knew the basics of chemistry but lacked knowledge about quantum mechanics and lasers, for example, that would be crucial in his future discoveries. He did make up for these deficiencies in graduate school. America was a culture shock after his Middle Eastern life. For him, friends had always been very important; from friends he could borrow money without writing it on paper and he could visit them without calling first. In case of a crisis his friends would spend hours with him and talk to him. When a good friend left Alexandria to visit Cairo for a week, the others would see him off on his train. Soon after his arrival in Philadelphia in 1969, Zewail slipped in his light shoes on the snowy street and fell down. The cars passed by and people minded their own business. To him the contrast was telling. In Egypt, traffic would have stopped, a chair would have been offered to him in the middle of the street, and somebody would have rushed to him with mint tea. Upon his arrival at the University of Pennsylvania, Zewail gave his new professor a gift that his parents had selected and wrapped for him. His fellow students viewed this with suspicion, whereas to Zewail his teacher meant everything sacred. Sensing the unfriendly environment, he immersed himself in learning everything in sight and indulged in the exceptionally favorable conditions.

We conclude our collection with two examples from Russia. Petr Kapitsa (P78),[44] the young and promising Russian physicist, was sent on a mission by the Soviet authorities in the early 1920s as part of their quest to open up Russian science, and help it catch up with the West. When Rutherford did not want to take him into the Cavendish Laboratory in Cambridge, saying that with about 30 people around him he had no opening, Kapitsa pointed out that one additional person would be within the accepted error limit, and the clever reasoning persuaded Rutherford.

Kapitsa did his doctoral work in the Cavendish Laboratory, and was eventually elected as a Fellow of the Royal Society.[45]

Kapitsa used to spend extended summer holidays in Russia before returning to Cambridge, until in 1934 he was detained in Moscow. Following an initial period of depression, Kapitsa immersed himself in building up his institute in Moscow.[46] As in Cambridge before, he behaved somewhat autocratically; his staff respected him but were also afraid of him. He was brave and stood up even to Stalin. Kapitsa had a row with Beria, the head of the secret police, when the atomic bomb was discussed many years later. Kapitsa was on a high-level committee of top physicists, which was headed by Beria. At some stage Kapitsa compared Beria to the conductor of an orchestra who had the baton in his hand but had lost the score. Beria complained to Stalin that Kapitsa was a dangerous person. Stalin told Beria that he could dismiss Kapitsa but he was not to be touched otherwise. Kapitsa was sacked in 1946 but not arrested, and he moved to his dacha. He managed to set up a sort of laboratory there. Kapitsa and his two sons, who were then teenagers, managed to do good experiments and to publish important papers. After Stalin died in 1953 and Beria was executed, Kapitsa was restored to his former position and returned to normal life in 1954.

Lev Landau (P62) burst onto the physics scene at a young age. He graduated from the prestigious Leningrad Physical–Technical Institute in 1927, but had already entered graduate school a year before. During 1929 to 1931 he spent a year and a half in Western Europe working with some of the world's best physicists. Upon his return to the Soviet Union he first worked in Leningrad, then Kharkov, and finally in Moscow in Kapitsa's institute.[46] Whereas during his stay in the West he had appeared to be a communist, Landau gradually became disillusioned back home, although not with the basic ideals of socialism. In the early 1930s many scientists in the West also sympathized with the Soviet experiment. Landau was arrested in 1938, perhaps on suspicion of involvement in authoring or editing a pro-socialist but anti-Stalin leaflet. Landau behaved courageously in jail, maintaining silence for two months and declaring a hunger strike. Curiously, he was not accused of writing the anti-Stalin leaflet, but with activities that were aiming at wrecking the Kharkov institute where he had worked. He was released as a result of Kapitsa's protests, and the latter acted as guarantor of his good behavior. By the time Kapitsa fell from favor with the authorities, Landau had become indispensable to the Soviet nuclear bomb program. Landau felt himself a 'learned slave' and considered the Soviet regime to be fascist. As his participation was his shield from the authorities, after Stalin's death he no longer felt the necessity and quit the program.[47]

6

What turned you to science?

A wide variety of factors have drawn future Nobel laureates to science. My observations come mostly from my conversations with Nobel prize-winners of the last decades of the twentieth century, and some readings. The most frequent sources of inspiration were a book, a chemistry set[1] (and not only for future chemists) or other experimentation at home, and a teacher, especially a high school teacher. The home environment, family members, or a family friend are also important sources of inspiration. In many cases the various sources blended. Many future laureates took up science when they were in their early teenage years, while others first completed medical school, during or after which they were exposed to scientific research and changed from medicine to research. Music and some humanities were great competitors and what tipped the balance in most cases were a taste for research and a good mentor.

Books and chemistry sets

The most successful book in turning children to science has been Paul de Kruif's *Microbe Hunters*.[2] This book is about natural scientists and their quest for uncovering nature's secrets. It first appeared in 1926 and has remained in print ever since.[3] De Kruif was a PhD researcher in the bacteriology laboratory of the University of Michigan Medical School, after serving for two years in Europe during the First World War. He decided to try popular science writing in 1919 and moved full time to his new profession in 1922. By then he was at the Rockefeller Institute for Medical Research in New York. While there, he continued his research and published papers in immunology with John Northrop (C46). He also came across Jules Bordet (M19) and other important scientists. De Kruif helped Sinclair Lewis (Literature 1930) in writing *Arrowsmith*,[4] a story about Martin Arrowsmith, a tough young man hell-bent on becoming a microbe hunter. *Microbe Hunters* was a decisive influence on, among others, Paul Berg (C80), Gertrude Elion (M88), Carleton Gajdusek (M76), Aaron Klug (C82), Leon Lederman (P88), César Milstein (M84), and Frederick Robbins (M54), scientists who were all born between 1916 and 1927. When he was 12, Gajdusek stenciled his 12 heroes from *Microbe Hunters* onto the steps leading to his chemistry laboratory in the family attic. Decades later, when he was a Nobel laureate and director of a large institute at Stanford University, Berg gave a copy of *Microbe Hunters* to every student in his lab.[5] For Herbert Hauptman (C85),[6] the earliest influences on his life were people that he read about, the philosopher Bertrand Russell (Literature 1950), Fermat, and Archimedes, and they became his role models. Bertrand Russell was also important for Edward Lewis (M95) in his formative years.

In this chapter the focus is on those first inspirations for embarking on science studies rather than on the motivation for choosing a Nobel Prize-winning research project. The latter comes up in other chapters. But here, again, the dividing lines are blurred in many cases.

Nikolai Semenov (C56)[7] read many books and his favorites were all chemistry books. He bought all the available chemicals in the nearest drug store and started experimenting with them. For him it was the greatest puzzle that sodium, this flammable and malleable metal, and chlorine, this extremely reactive gas, formed the innocent table salt. He burned a piece of sodium in chlorine gas and recrystallized the precipitate. It was a white powder, which he poured over a big slice of bread and it was indeed table salt. His interest in chemistry kept deepening until he read in a book that the future of chemistry was in physics and since his goal was to become a good chemist, he signed up for the school of mathematics and physics at St. Petersburg University.

William Lipscomb (C76)[8] grew up in Kentucky, where his father was a physician and his mother a music teacher. She gave him a chemistry set when he was 11 years old, and he started experimenting. He discovered that he could buy additional apparatus and chemicals at a special rate, using his father's privilege at the drug store. William, who had friends who were also interested in science, went to a high school that had rather poor facilities, and his father threatened to take his son to another school unless they introduced chemistry. They did, but after the first chemistry class the teacher asked Lipscomb to come back only for the examination because he already knew everything. He went through college on a music scholarship at the University of Kentucky, and had a separate program for things that were not taught. For example, nobody in chemistry knew anything about quantum mechanics, and he studied it on his own.

Paul Boyer (C97)[9] at the age of ten and Robert Curl (C96)[10] at the age of nine were similarly first exposed to science through a chemistry set. George Olah (C94)[11] and a friend set up some chemistry experiments in the basement of the friend's house. It ended up in a small explosion and disastrous stink. His friend's parents closed down their 'lab'. Olah finally fell in love with chemistry when he started university. It impressed him with its practical aspects of making materials, plastics, and pharmaceuticals —all man-made compounds and synthetic materials essential to modern life. Leon Lederman (P88),[12] too, was interested in chemistry when he entered City College in New York. In addition to his readings and good chemistry teachers in high school he also had a small chemistry laboratory at home. By the time Lederman finished college, his interest had shifted to physics. He found physics easier and organic chemistry discouraging. He also found the physics majors more interesting than the chemistry majors. College chemistry also discouraged John Vane (M82)[13] because it was all recipes. His interest in science had begun when his parents gave him a chemistry set at the age of 12. He did his experiments in their kitchen until an explosion prompted his father to build a garden shed for him. His chemistry teacher further enhanced his interest. Following college he opted for pharmacology. A chemist aunt encouraged Mario Molina (C95) to conduct chemical experiments. He grew up in Mexico, and rather than having a traditional career in engineering, he decided to become a research scientist. He studied in Mexico, which was easy, and in Germany where he felt no pressure, but when he got to the University of California at Berkeley, he had to work hard. Apparently the challenge encouraged him.[14]

George Porter (C67)[15] received his first chemistry set when he was ten and started doing science. He could do marvels with it in the kitchen and produced his own fireworks. It was a challenge to Porter that his father was a Methodist preacher and his son had to attend service. He asked awkward questions and read Thomas Paine and others who were questioning religion. This experience also strengthened his interest in science. John Walker (C97)[16] faced a similar dilemma. He was brought up to be religious but then rejected it because for him it was either science or religion. Walker was interested in modern languages, but there was a young chemistry teacher who was very stimulating so he opted for maths and science. Walker studied chemistry in Oxford where Cyril Hinshelwood (C56) gave the very first lecture he attended. During his undergraduate period he became rather disillusioned with chemistry and his attention gradually shifted to biochemistry. Religion, or rather its rejection, also played a role in James Watson's (M62) interest in science, in which his father fully supported him. Watson loved the outdoors and intended to be a birdwatcher, acquiring a book about bird migration when he was seven years old. His father had been a birdwatcher for years. It was only during his university studies that his interest shifted to molecular biology. 'I was curious of why the world is like it is? Laws of nature. Why did things happen?'[17] When Watson got a little older, the question 'What is Life?' seemed to be paramount.

Teachers

Jean-Marie Lehn (C87) also did chemical experiments at home and the first one ended in an explosion! It taught him that he had to learn to survive. He also loved music. A primary school teacher, very dedicated and wanting to promote his pupils, was an especially strong influence on Lehn's development. He worked with them overtime to prepare them for high school. Then, in high school, in Lehn's words,

> I wasn't sure I liked or not what I was doing but when I began to read philosophy things started to change. At 15 or 16 you begin to ask yourself questions, and you question things that you had accepted before. As Paul Valéry said, 'You think as you feel resistance.' You bump into something, and you begin to think. This is exactly what happened to me.[18]

During the first decades of the twentieth century a strong high school system in Hungary, and especially in Budapest, encouraged several future Nobel laureates and other great scientists to become interested in science and mathematics. They included Eugene Wigner (P63), Dennis Gabor (P71), George Hevesy (C43), George Olah (C94), Albert Szent-Györgyi (M37), and John Harsányi (Economics 1994), and Theodore Kármán, John Neumann, Michael Polanyi, Leo Szilard, and Edward Teller. It was a juxtaposition of a good system of education and a mostly Jewish upper middle class, in which advancement was possible through learning.[19] Wigner[20] had an exceptional maths teacher in the Lutheran high school in Budapest,[21] who helped him at an early age to learn more about his favorite subjects and gave him books to read.[22] (Wigner remembered him in his speech at the Stockholm City Hall during the Nobel week in December 1963.[23] He even had a word for his extraordinary schoolmate, Neumann, in the same speech.)

Wigner was so much taken by the books that he wanted to study physics after high school. At that time his father asked him, 'How many jobs for physicists are there in our country?' That was an embarrassing question. Wigner exaggerated a little and said, 'Four.' A couple of years later the same thing happened to Gabor, whose answer to his father's question was 'Six.'[24] Both Wigner and Gabor then studied engineering at the Berlin University of Technology; Wigner studied chemical engineering and Gabor electrical engineering. However, both sneaked over to Berlin University, whenever they could, to attend its famous seminars in physics with the great Albert Einstein (P21), Max Planck (P18), and others.

Arthur Kornberg (M59), Jerome Karle (C85), and Paul Berg (C80)[5] all went to the Abraham Lincoln High School in Brooklyn. There were special classes for gifted students and they could combine two years into one. There were extraordinary groups of young people in this school, who fed on each other and the level of excitement in learning, and did projects that were not in the regular class program. The school provided a strong motivation for science. It was the job of Sophie Wolfe, who supervised the school stockroom, to supply the various classes in chemistry, physics, and biology. She was not a teacher but she loved the students and started science clubs. The students stayed after classes were done for the day and worked in her laboratory. She never gave an answer to a question but always encouraged the students to find out for themselves. Sometimes that meant doing an experiment, sometimes it meant going to the library, but it was always the student who had to solve the problem. After Berg's Nobel Prize, *The New York Times* ran a full-page article about her and some years ago the city of New York named one of the wings of Abraham Lincoln High School the Sophie Wolfe Wing.

When Berg came back from the navy after the war, he took his undergraduate degree in 1948 at Pennsylvania State University. By that time, because of his summer-job experience, he knew he wanted to proceed to a PhD. That experience had been acquired at the Lipton Tea Company and at General Foods, doing analytical chemistry in the lab. There were people in these labs who made the decisions about what experiments others were to do. They had PhD degrees and the ones who had to do the experiments had bachelor's degrees. Berg wanted to be the one who decides what experiment is to be done.

His high school in California was decisive in turning Glenn Seaborg (C51)[25] to science. Until his junior year Seaborg was uninterested in science, but when he took a course in chemistry a miracle happened; he was turned on by an inspiring high school teacher. He told the students about interesting work that was going on, controversies, and the big discoveries that were being made. Seaborg decided to become a chemist. Later he went to Berkeley for graduate work, and he did it in nuclear physics. For John Cornforth (C75),[26] who grew up in Australia, high school experience was also decisive. His choice of science was conditioned by deafness, since he had to find something in which it would not be a fatal drawback. He had a good chemistry teacher, and began to be interested in the subject.

One more example of high school-teacher influence is James Black's (M88).[27] His intellectual awakening began at the Cowdenbeath Secondary School, which was the main school in the local mining community. He came under the influence of Dr. Waterson, the mathematics teacher, who, like Black, had grown up in the culture of the east Fife coalfields in Scotland. Education, leading to self-improvement and escape from the dangers of mining, was the most important cultural activity in that community. Waterson grew up there and went to St. Andrews University in Fife to study mathematics. He was a brilliant student but did not take up a career in academic research because he was programmed by his upbringing to become a schoolteacher. He earned his PhD and then DSc while still a full-time schoolmaster. When Black was 14 years of age, Waterson gave him a copy of *Calculus Made Easy* to work through on his own. The answers to problems were at the back of the book, and when more than halfway through the sections on integral calculus, Black got a different answer to a problem from the one in the book. When he asked Waterson for help, the teacher worked through the problem and reached the same answer as his pupil. His response was astonishing: 'The book is wrong.' He had the knowledge and confidence to challenge the authority of the book; this was Black's epiphany and he has been irreverent in his thinking ever since. Waterson persuaded Black to sit the scholarship examination for St. Andrews University and, at 16, he was offered a residential scholarship. Black graduated in medicine in 1946 and started a career in physiology at the medical school in Dundee right away.

Learning irreverence to authority was an important component of Black's education. Herbert Brown (C79)[28] had some similar experience when he was doing chemical analysis as a college student. First Brown studied electrical engineering, and chemistry, which was a required subject, fascinated him. However, they closed his college for lack of funding. One of Brown's former teachers, Nicholas D. Cheronis, originally from Greece, ran a small commercial laboratory in his home, called 'Synthetical Chemicals', as a side business. He invited Brown and other students to spend their time in his laboratory. Brown also registered for a correspondence course by Professor Julius Stieglitz on qualitative analysis at the University of Chicago. They sent him the unknowns; he did the analysis in the laboratory, and sent the reports in. One time Brown got an unknown that just did not make sense; it did not behave the way it should. It turned out that the unknown contained something that it was not supposed to contain. Stieglitz acquiesced and asked his assistant to send Brown a new sample. This encounter with Stieglitz was the beginning of a relationship that changed Brown's life. When, eventually, Brown became a student of the University of Chicago, Stieglitz talked him into going to graduate school. Another influence was when his newlywed wife bought him a book, *Hydrides of Boron and Silicon*, by Alfred Stock. It sparked his interest and he started to do graduate work for H. I. Schlesinger, who was one of the two principal world experts on boron chemistry at that time.

In addition to Brown, Sune Bergström (M82),[29] Kenichi Fukui (C81),[30] and Aaron Klug (C82)[31] provide examples of college teachers' beneficial influence in turning future Nobel laureates' attention to science. Bergström grew up in Stockholm. His entry into science was accidental, but his first teacher was a very stimulating Finnish chemist, Erik Jorpes. He worked on heparin so Bergström worked on it for a couple of years, too. Jorpes had royalty income from heparin and used some of that money to send Bergström to study in London. Fukui grew up in the ancient Japanese city of Nara. He was in the middle school when he became interested in nature and joined a biology-oriented group of pupils. He read *Souvenirs entomologiques* by Jean Henri Fabre and Jules-Henri Poincaré's trilogy of science-philosophy. Chemistry was not his favorite subject; he had the impression that it was lacking in logic. Then a chemistry professor at Kyoto University explained to him 'the hidden logical character implicit in chemistry', and Fukui decided to make it his field. Klug started in medicine at the University of Witwatersrand in Johannesburg. Gradually he became more interested in biochemistry, giving up anatomy but continuing with physiology, biochemistry, and histology. Eventually he switched from medicine to science for his BSc degree. Then he took a MSc degree in physics at the University of Cape Town under Professor R. W. James. James had been a colleague of Lawrence Bragg (P15) in Manchester. Then Klug went to Cambridge, because Cambridge was the place to go, and his interest in science became firm and final.

Whereas Klug changed to science during his medical studies, others first completed medical school and switched later. These included François Jacob (M65), Gerald Edelman (M72), Daniel Nathans (M78), and Günter Blobel (M99). The research atmosphere in André Lwoff's lab at the Pasteur Institute in Paris infected Jacob. Edelman did some research with Britton Chance that had a long-ranging impact. For Nathans the influence came from summer research with Oliver Lowry, professor of pharmacology at Washington University in St Louis. They worked side-by-side at the bench and the experience completely changed Nathans' plans for the future. To Blobel research meant a whole universe of the unknown and he realized that if he were to go into medical practice, he would never be able to explore it.

Joshua Lederberg (M58)[32] went to Columbia College for premedical training. During his summers he worked as a parasitologist in a hospital and this further strengthened his interest in science rather than in practicing medicine. Lederberg had been a student of the Stuyvesant High School in New York City, which focused on science and jumpstarted the careers of several Nobel laureates. One of those was Roald Hoffmann (C81),[33] for whom the social setting of a small research group, the closeness to science and scientists, and the turning into reality of what was taught in courses, was important. He decided to become a chemist when he was about 20 years old.

Edward Lewis (M95)[34] grew up in Wilkes-Barre, Pennsylvania. He read H. S. Jennings' *The Biological Basis of Human Nature*, which had simple diagrams explaining Morgan's law of sex-linked inheritance. In *Science* magazine he saw an advertisement for cultures of *Drosophila* that could be ordered from a professor at Purdue University for $1 each. With a friend Lewis ordered cultures and began growing *Drosophila* in their high school biology laboratory, and it has remained the center of his entire research career.

Family

Family background, with its direct and indirect influences, has played a decisive role in many future Nobel laureates' lives. The beneficial academic families of John Polanyi, Kenneth Wilson, and Alfred Gilman were mentioned at the beginning of the previous chapter, as was that of Carleton Gajdusek[35] for whom his aunt, Irene Dobroczki, made all the difference. An entomologist in plant diseases and pioneer in insect tissue cultures, Dobroczki was one of the first women in *American Men of Science*. She was also a friend of Barbara McClintock (M83). When Gajdusek was about five or six, his aunt hauled him along to her institute and to the Rockefeller Institute for Plant Research at Princeton University and other laboratories. She introduced him to artists, poets, writers, and play actors in post-World War One New York City, and to prominent scientists in virology, plant pathology, genetics, and entomology. He was thinking of going into medicine and William J. Youden, who was professor of mathematics at Columbia University, advised him to study physics and maths. Once Gajdusek asked him, in the greenhouse, what would happen if instead of sunlight they were to use different colors of light, and Youden let him run his own experiments there.

One of Richard Smalley's (C96)[36] family influences was also an aunt, Sara Jane Rhoads, one of the first female professors of chemistry in the United States. Smalley suddenly became serious in his studies at some point in high school, first in chemistry, later in physics. This was the era when sputnik had made the careers of engineers and scientists romantic for young boys. Smalley's father had a woodworking shop in the basement of their house, and from him he learned to tear things apart, figure them out, and put them back together. They built many things together. While Smalley was in high school, his mother studied part-time to finish her liberal arts college degree and he watched her fall in love with knowledge. She taught him to do mechanical drawing in those early pre-chemistry days when he thought he was going to be an architect.

Rocket fuels turned Kary Mullis (C93)[37] to science. He was already doing experiments at home at the age of 13. He could buy a pound of potassium nitrate in the drug store and a hundred feet of dynamite fuse in the hardware store, with no questions asked. This was in the late 1950s and Mullis was excited by the sputniks. He built a rocket of aluminum pipe, four feet tall, with an asbestos nozzle impregnated with plaster of Paris holding it together, and put a tiny frog inside an aluminum film canister in the top of his rocket. The rocket went high and brought the passenger back alive.

Family influence may come in very different ways. For Gertrude Elion (M88)[38] the death of her grandfather from cancer was a strong motivation in becoming a scientist. This happened when Elion was 15 and about to enter college. She wanted to study to find a cure for cancer. Joshua Lederberg's father was a rabbi and there were other rabbis among his ancestors, who stressed the importance of learning. In this he followed the family traditions but he chose science for his actual inquiries.

Emilio Segrè (P59)[39] was born near Rome. He had two uncles, a jurist and a geologist, and both were members of the Academy of Science. It was an intellectual environment. As soon as Segrè learned to read he became interested in physics because he found physics books at home; there was even one with experiments for children. He started doing these experiments and described them in his own notebook with drawings. When the family moved to Rome in 1917, Segrè went to school, learning Latin, Greek, and other conventional subjects. When one of his uncles gave him a physics book, he inscribed it to him with the dedication that he hoped physics would be used for peace. The First World War was still raging.

Owen Chamberlain (P59),[40] whose father was a radiologist, lived with his family in Philadelphia. A family friend, an amateur magician, played fascinating number games with the 12-year-old Owen, and explained to him why it is impossible to make a perpetual motion machine. Unbeknown to Chamberlain, the man was a physicist.

Ilya Prigogine's (C77)[41] family emigrated from Russia to Belgium when he was small. He had a brother who had studied chemistry before being drawn to African birds and becoming a well-known ornithologist. Their father was a chemical engineer. Prigogine started by studying archeology and history, and played a lot of music. Then he became interested in philosophy and read Heidegger, Whitehead, and Bergson. Everywhere he encountered the problem of time. However, when he entered university in the years before the Second World War, everybody was saying that it was not the right time to become a philosopher. Following the chemists in the family, he went into chemistry and took up physics at the same time. He was surprised to find out that time meant very different things in philosophy and science. In philosophy it was considered to be a very difficult question of ethics and human responsibility, 'our basic existential dimension'. However, his physics teachers explained to him that the problem of time had been solved by Newton, and completed by Einstein, and there was nothing to look for anymore. Prigogine decided to devote his life to understanding and resolving this contradiction.

Frederick Sanger (C58, C80)[42] originally expected to become a doctor, like his father. However, his father was always going from one patient to another, and Sanger, feeling that he would be much better at something where he could really get to work on a problem, decided to study science rather than medicine. His brother was the first important influence on him: they were close in age, but to Sanger he seemed considerably older since he was the leader. The introverted Frederick used to follow his extrovert brother around, and learned a lot from him. The brother was interested in nature, particularly in snakes, and was an expert at finding birds' nests. Eventually he became a farmer. Their father was also interested in science and nature, and did some work in Cambridge, in what is now the Molteno Institute, identifying bloods by immunological techniques. He also spent some time with Scotland Yard showing them how to identify human blood. Thus he was an influence, too.

Johann Deisenhofer (C88)[43] grew up in a village in West Germany. His father ran the family farm, but his interest was much broader than the farm and he transmitted this to his son who became an avid reader of popular science books. The county library sent a bus with books to their village weekly. Deisenhofer's interest in physics started with a book by the Cambridge astronomer, Fred Hoyle.

Marshall Nirenberg (M68)[44] lived with his family in the outskirts of Orlando which, before Disney World, was a small town. Florida was unspoiled at the time, and they lived within half a mile of a tremendous swamp that went for 20 miles without a house. Nirenberg was interested in biology and collected spiders for the American Museum of Natural History. When he went to camp as a child, he always won the nature award. There was a big airbase at Orlando, which ran a jungle survival course for pilots who were going to the South Pacific. They had collected professional biologists from universities and museums all over the United States to teach this course. The ornithology professor invited Nirenberg and his two friends to a birding expedition. Nirenberg taught comparative anatomy as an undergraduate and when he went to graduate school he had a part-time job as a technician in the nutrition laboratory, working with radioactivity. This was his first exposure to biochemistry.

Ahmed Zewail (C99)[45] went to school in Alexandria, Egypt and achieved high grades in science. During his high school years he set up experiments at home. Intrigued by the changes that substances underwent in transforming from solids to burning gases, Zewail would heat material like wood in test tubes. Fortunately, there were no explosions. Learning was everything for Zewail, and when he went to visit the University of Alexandria for the first time and saw its beautiful, ornate buildings, the shrine of knowledge and learning, it brought tears to his eyes. By then he knew that he was going to become a scientist.

7
Venue

A decisive influence for a research career is for it to be launched in a strong environment. The adviser counts the most, but the whole atmosphere is important, the other professors and fellow students, the technological level of the institution, the visitors, and so on. The research seminar is probably the single most critical ingredient in shaping the young researcher's career. It broadens his horizon, introduces him to new fields and outside scientists, with different styles and approaches, and teaches him how to conduct scientific discussions. The beginner sees how questions are asked and answered, witnesses the debates, and gradually becomes part of the process. A diverse selection of examples, of both great venues and lesser-known ones, will come up in our discussion, often as seen through the eyes of the Nobel laureates, looking back.

It would be rewarding to find a general recipe for producing venues that best nurture Nobel-level work and Nobel Prizes. Alas, this is impossible. It may, however, be possible to discern some characteristic features. Much depends on the individuals and the atmosphere they create. Freedom, democracy, and an informality in interactions are assets of a good venue. This does not mean that scientific questions can be decided by popular vote. Ideology is best left out of the laboratory. The communist J. D. Bernal created a successful venue of crystallography for molecular biology and materials science within the Department of Physics at Birkbeck College in London, which had its best times in the 1950s. When he was asked whether he ran his lab on communist lines, he answered that he had only advanced as far as feudalism: his co-workers ploughed the lord's land for half the time and for the other half they cultivated their own patches.[1]

An inspiring environment in his formative years helps a young researcher to develop his own ideas. In James Watson's (M62) set of rules about succeeding in science, the first one is 'to avoid dumb people' and the second is 'you must always turn to people who are brighter than yourself'.[2] Carleton Gajdusek (M76) recognized this, too: 'From my late teens I understood that bright people hang out with other bright people.'[3] Having completed his MD degree and his residency in New York, Gajdusek went to Cal Tech: 'Aage Bohr, Jack Dunitz, Benoit Mandelbrot, Gunther Stent, Jim Watson, Ellie Wollman, Wolf Weidel, and many others were in our group. We were greatly influenced by Linus Pauling, John Kirkwood, Max Delbrück, George Beadle, James Bonner, and Laszlo Zechmeister.'

As important as an excellent venue is for a budding scientist, there is merit in not being initiated at the foremost institution, if the person is strong enough to turn this to his advantage. James Watson sees such advantage in his having grown up in Chicago, and going to the University of Chicago rather than to Harvard. 'I saw no reason to treat authority with much reverence. You were never held back by manners, and crap was best called crap. Offending somebody was always preferable to avoiding the truth . . .'[4] Watson then got his second chance of avoiding Harvard when it did not take him for graduate studies, and neither did Cal Tech.[5] Instead, he ended up at Indiana University in Bloomington. H. J. Muller (M46) was there, recently awarded the Nobel Prize 'for the discovery of the production of mutations by means of X-ray irradiation'. More important for Watson, Salvador Luria (M69) and Renato Dulbecco (M75), two future Nobel laureates, were also there, and another, Max Delbrück (M69), occasionally visited. Delbrück's name was familiar to Watson from the pages of Erwin Schrödinger's (P33) *What Is Life?* Indiana was coupled with the summer meetings at Cold Spring Harbor, which was also an important venue (see below) in many research careers.

It may be instructive to also consider a negative example, to illuminate the importance of the intellectual atmosphere of a good venue for scientific research. Between 1947 and 1950, James Black (M88)[6] worked in the medical school in Singapore. He taught and did research in physiology. He made new equipment for measuring blood pressure and blood flow so he could study the pressure–flow relations of the intestinal circulation. Although he obtained many results, he found it hard to judge their value. He had nobody to talk to about his work, and oscillated between thinking 'this is great' and 'this is rubbish'. There he learned the tremendous importance of contacts with colleagues, and how vital it is to normalize one's intellectual activity.

Ingredients of success

One of the great venue success stories is the Laboratory of Molecular Biology in Cambridge. Many future Nobel laureates spent shorter or longer periods of time there, and the following had close associations with it: F. Sanger (C58, C80), M. Perutz (C62), J. Kendrew (C62), F. Crick (M62), J. Watson (M62), A. Klug (C82), C. Milstein (M84), G. Köhler (M84), and J. Walker (C97).

It all started in a humble way. When Max Perutz's funding at the Cavendish Laboratory was running out in 1947, he needed financial support, which had come first from the Rockefeller Foundation, later from the Imperial Chemical Industries, though, curiously, never from Cambridge University. David Keilin, the great Cambridge professor of biology, suggested to Lawrence Bragg, who was director of the Cavendish Laboratory, that he turn for support to the Medical Research Council (MRC). Perutz by then had been working on the structure of hemoglobin for about a decade, and it did not at the time seem to be a very promising project for the MRC. From a luncheon at the Atheneum Club between Bragg and Edward Mellanby, the MRC secretary,[7] the MRC Research Unit was born within the framework of the Cavendish Laboratory. Eventually, in 1962, the unit moved to new premises, and has operated since as the MRC Laboratory of Molecular Biology (LMB).

Even today, when over 400 people work in the LMB, its director can paint it in very attractive terms:

> We have a kind of federal system in which everybody has some
> influence and some ownership in the Lab as a place to work. It is
> not a communist system, but there are other ways to distribute the
> power. We do not have much grant proposal writing to do, we have
> no teaching to do, and a minimal amount of administration.[8]

At some point James Watson (M62)[9] made a critical remark about the superior research conditions at the LMB, which hinder its associates from moving out and establishing centers of excellence elsewhere. The practice of sending away excellent people to start off new centers of research was a Bragg legacy at the MRC LMB. The elder Bragg practiced it when he was at the Royal Institution. In the late 1920s he suggested to one of his associates, William Astbury, that he should move to Leeds. Astbury did so, and established the world's leading center of fiber research. William Bragg's way of directing his associates' attention to various fields was indirect and subtle.[10] He would just ask people like Astbury to help him with photographs to illustrate his lectures for children. This is how Astbury became interested in the fibrous state, a fascination that stayed with him for his entire career.

Three quotes follow, from conversations about the LMB. One is by the former American postdoctoral fellow Sidney Altman (C89), who became a professor at Yale University and won the Nobel Prize in 1989. The second is by Michael Fuller, the former laboratory steward, now retired. The third is by John Walker (C97), who spent his more senior period of research at the LMB.

Sidney Altman did a postdoctoral stint in 1969/70 at the LMB, during which he started working on transfer ribonucleic acid (tRNA). He said:

> Having grown up wanting to become a physicist, I'd read a lot
> about the history of 20th century physics. I thought that to be
> in a place like Copenhagen in the 1920s and '30s must have been
> amazing. I regarded the MRC Lab in Cambridge in the same way
> because in my field at that time it was the pre-eminent lab in the
> world. It seemed as if major discoveries were being made there
> every few months. I was overjoyed to have the opportunity to go
> there. I knew it would be a great experience. What was great about
> it was that there were about a dozen absolutely amazing scientists
> there, several of whom had already won the Nobel Prize. They set
> the tone for how the place worked.
>
> Some of the individuals I alluded to included Perutz, who was
> the head of the lab, Brenner, Crick, Sanger, César Milstein, Hugh
> Huxley, and Aaron Klug. Kendrew wasn't around very much by
> then. But the others were working every day in the lab, alongside
> everybody else. Fred Sanger was a great ultimate example. He,
> himself, was working on DNA sequencing and he only had one or
> two people working with him on the project. He was synthesizing
> compounds himself.
>
> Everybody went to tea, according to the English custom,
> mid-morning and mid-afternoon. These 'gods' of molecular
> biology were there, sitting with everyone at tea. They encouraged

everybody to participate freely in discussion and they treated everybody equally. You could be the youngest graduate student or a technician, it didn't matter. They took your ideas seriously. They were also honest and very forthright in their criticism.

You had to develop yourself in two ways. On the one hand, you had to train yourself not to say anything that was superfluous or stupid. The 'gods' made it very clear if they felt that you were frivolous or you were thinking in a less than coherent way. One could become very discouraged very quickly if that happened. Some people succumbed. On the other hand, if you were able to engage in the discussions, take advice, and approach the gods only when you had really something to say, it was fantastic. They made you feel part of the enterprise.

Crick and Brenner shared an office, which was highly unusual. They were generating new ideas all the time and found it important to bounce their thoughts off each other. Their door was always open to others. There was a sign on it saying something like 'If this door is closed, just knock and walk in if you are member of the lab. If you are not a member of the lab, please speak to one of the secretaries first.' But the door was open 95% of the time. The same was true with all the senior scientists. The most important thing for a young scientist was learning to engage in discussions with anyone and to be aware that once you were engaged in a discussion, you had to be rigorous in your thinking and had to be able to accept criticism without taking it personally.[11]

It may have been the Cambridge influence that showed later in the economy of interaction in James Watson's laboratory at Harvard University. Benno Müller-Hill remembers:

there was Jim Watson, who didn't talk at all. When he came into the lab, you would say one or two sentences, and that was it. You just realized that there was nothing more to say. For example, once I got a particular mutant and I was very proud of that but you can explain such an experiment in two minutes. There was another German postdoc there whom I knew from Freiburg, Klaus Weber. We had a bet about who spoke how much with Jim. After half a year, he was at 22 minutes and I was at 17 minutes, total. Jim Watson's non-speaking was even more driving. It had the effect that there was someone with whom you could speak only if you had something to say. If you had no result, there was nothing to speak about.[12]

Michael Fuller looked at the LMB from a different perspective. He was born in Cambridge and went through the local grammar school, which specialized in technical education, training students in the use of machine tools, and typing, commerce, etc. Fuller had a photographic memory, and later on he would read the advertisements in *Nature* and could remember what was where when the lab needed something special. Fuller's mother worked as a cleaning lady in the university's Department of Biology. Michael had an interest in science, and had always wanted to work at the Cavendish Laboratory. When his formal education ended at 15, he wrote to Sir Lawrence Bragg asking for a position there. A group of three future Nobel laureates, Perutz, Kendrew, and Crick, plus the lab engineer, invited Fuller for an interview. Obviously, a great deal of attention was paid to who was hired, even for such a junior position. It was not unprecedented either: when Ernest Rutherford (Co8) was to assume the Cavendish professorship in 1919, he interviewed and hired G. R. Growe for an assistant's position. When Growe retired in 1959 Cambridge University admitted him to the degree of MA in recognition of his dedicated work.[13] Here are Fuller's memories of his job interview and his impressions of LMB:

> I was being asked various questions, I remember the one about what is sodium chloride and I told them it was salt, and I was hired. Initially the job was as an apprentice technician in physics. I had a good grounding in physics at the Cavendish. It was very much sealing wax and string. You made your equipment yourself and got it running. I also learned that there's no such word as 'can't'. Somehow you got the work done. We worked very long hours, initially just getting the rotating X-ray sets working. At 23 I became the lab chief technician, in 1959, and I was lab steward, superintendent, lab manager all in one person. I reported directly to Dr. Perutz, the lab director. I thoroughly enjoyed it. In those days we had about 70 people. We set up a small canteen. Mrs. Perutz took charge of that. It was very interesting because in 1961–62 it was the first canteen in the country where all members of the staff queued to have a cup of tea or sandwiches. There was no discrimination between scientists, technicians, and secretarial staff. It actually made front page in *The Times*, saying, 'What on earth's happening in Cambridge?' It really had the effect of bringing the entire lab together. Everybody was treated as equal and I think it was one of the secrets of the lab, the collaboration, the fact that the virologists would sit at the same table as the

computer engineer and discuss problems. Looking back, it was so revolutionary at the time whereas now it's so standard. The canteen was almost the most important room in our building.[14]

John Walker (C97) comments on the question of how it is possible to maintain the high level of research for an extended period of time:

> We have had a long period of very good leadership, from people like Perutz and Sanger, by example rather than by political decision. We have a culture here and we are sufficiently mature to realize that if the place is going to stay at the top it has to be renewed from the bottom. This is why we spend a lot of time over recruitment. We know we have to replace ourselves. Otherwise the place goes into a downward spiral from which it would probably never recover. Until recently I was chairman of the recruitment committee. The policy has been, find good people and let them do what they want to do. This is how I got into ATP, which was a very unfashionable topic to work on at that time. But I did receive support from Fred [Sanger] because he thought that it was great because it was so difficult and he told me so, 'If it is difficult, go for it.'[15, 16]

Power of tradition

The Cavendish Laboratory in Cambridge was one of the roots of molecular biology. The director of the Cavendish Laboratory, Ernest Rutherford (C08), died in 1937. Under him the laboratory had enjoyed tremendous success in modern physics. Lawrence Bragg (P15) became his successor. Curiously, Rutherford was a chemistry Nobel laureate and Bragg a physics laureate. Very soon after Bragg assumed the directorship the Second World War began, and with its completion hard decisions had to be made about the future of the laboratory. Atomic physics had become a big science, one which the Cavendish was ill-equipped to pursue. 'Rather than fight a rearguard action',[17] Bragg's support of molecular biology and radio-astronomy enabled the Cavendish 'again to lead the world'.

Antony Hewish (P74) discovered, in his radio-astronomy research, that the upper atmosphere, the ionosphere, was affecting radio waves on their way through from space. The ionosphere, which is at a height of about 300 km, was discovered at the Cavendish by Edward Appleton (P47) before the Second World War. It resembles a layer-like glass sphere, and the variations in its thickness cause random diffraction of radio waves, which can be picked up on the ground by fluctuations in intensity. Hewish found it appealing that one could measure something about the ionosphere using radio waves from space. He was intrigued by the possibility of using radio-astronomy to study the upper layers of the ionosphere. This was in the 1950s, before the direct measurements that would become possible with space research. In this work, the sources of radio waves were of no interest. Later research brought spectacular results in this respect, the discovery of pulsars.

Hewish graduated in 1948 and joined Martin Ryle's (P74) research group at the Cavendish Laboratory. Ryle and his colleagues developed the method of the so-called aperture synthesis. Instead of making one huge aerial, a number of small aerials were used and the signals received by them combined, leading to a resolution corresponding to one aerial of enormous size. With this approach, Ryle could map cosmic radio sources with great accuracy. Bragg encouraged and supported the work. He recognized the similarity between Ryle's approach and some of the techniques used in X-ray crystallography.[18]

W. L. Bragg's task as director in keeping the Cavendish Laboratory in the forefront of science was not trivial. The roster of his predecessors proclaimed some important eras. These included James C. Maxwell, famous for his studies of electromagnetic waves and his equations; John W. Strutt, better known as Lord Rayleigh (P04), the discoverer of noble gases; Joseph J. Thomson (P06), the discoverer of the electron; and Ernest Rutherford (C08), the discoverer of the atom's nucleus. Originally from New Zealand, Rutherford came to the Cavendish Laboratory after he had been professor in Montreal and later in Manchester. In Montreal he had other shining names as associates, such as Frederick Soddy (C21)[19] and Otto Hahn (C44). The young Dane, Niels Bohr (P22), joined him in Manchester after not hitting it off with Thomson in Cambridge, and so did George Hevesy (C43).

Rutherford took over the Cavendish from Thomson in 1919. Under Rutherford the laboratory probably had the largest concentration of present and future Nobel laureates, including Edward Appleton (P47), Francis Aston (C22), Patrick Blackett (P48), James Chadwick (P35), John Cockroft (P51), Paul Dirac (P33), Petr Kapitsa (P78), Nevil Mott (P77), Joseph Thomson (P06), Ernest Walton (P51), Charles Wilson (P27), and Rutherford himself (C08). Although Rutherford was approachable, many students hesitated to ask him questions. There were always others to turn to, especially James Chadwick.[20] Whereas Rutherford 'always wanted to tell you what he wanted to say and not what you wanted him say, . . . Chadwick was a simpler person. One could understand his thinking much better than one could understand Rutherford.'[21] According to Chadwick, it was under Rutherford at the Cavendish that, for the first time, 'a great laboratory had concentrated so large a part of its effort on one particular problem',[22] that is, on the atomic nucleus.

Ernest Lawrence's (P39) Berkeley laboratory showed a similar concentration on one particular problem, particle accelerators and radioisotopes. He was awarded the Nobel Prize for the cyclotron, but it 'honored not only the invention of an instrument but also the creation of the environment necessary to exploit it'.[23] Lawrence came to Berkeley from Yale and 'by 1930 he had more graduate students than any other member of the Physics Department'.[24] Several Nobel laureates emerged from Lawrence's circle, including Edwin McMillan (C51), Glenn Seaborg (C51), Emilio Segrè (P59), Owen Chamberlain (P59), and Luis Alvarez (P68). Mark Oliphant, who played a crucial role in maintaining the British–American connection with the atomic bomb project during the Second World War found Lawrence's Berkeley lab like Cambridge: 'It was a place where everybody interested in nuclear physics went.'[21]

It is a true sign of success when something gets imitated and when Petr Kapitsa organized his new Institute of Physical Problems in Moscow, in the 1930s, this is what he was trying to do, and his model was Rutherford's way of running the Cavendish.[25] Thus, for instance, the laboratory work stopped at a certain hour in the evening to let research workers have time to reflect. Kapitsa insisted on occasions that his subordinates must take a break, whether it suited them or not. He further insisted on knowing everything that was going on in the institute, and approving any new apparatus before it was made in the workshop.

It was true that Rutherford did not like his co-workers staying in the lab too late into the night. However, he did not hesitate to call Mark Oliphant at 3 a.m. when he had understood their experiment of the previous day. It was about the discovery of helium-3 along with tritium.[21] Rutherford was the professor and Oliphant a student; the early morning call would not have happened the other way round. When, almost two decades later, under Lawrence Bragg's directorship Max Perutz did a crucial experiment on a Saturday, providing direct evidence for Linus Pauling's α-helix, it was not until Monday morning that he 'rushed' into Bragg's office to tell him what he had discovered. He would not have disturbed Bragg at home over the weekend.[26]

In the 1920s in Copenhagen, Bohr built up one of the great centers of theoretical physics of all time. His co-workers, for longer or shorter periods, included the future Nobel laureates Paul Dirac (P33), Werner Heisenberg (P32), George Hevesy (C43), Lev Landau (P62), Nevil Mott (P77), Wolfgang Pauli (P45), and Harold Urey (C34). Other centers of physics included Arnold Sommerfeld's in Munich, Max Born's (P54) in Göttingen, Erwin Schrödinger's in Zurich, and, of course, that at Berlin University, with its famous Thursday afternoon colloquia with such giants as Albert Einstein (P21), Max Planck (P18), Max von Laue (P12), and Walther Nernst (C20) in the front row, and 'youngsters' such as Dennis Gabor (P71), Heisenberg, Pauli, and Eugene Wigner (P63) in the back.

A symposium venue

Molecular biology had its beginnings in the 1930s, and the Cold Spring Harbor symposia provided an important venue for it. These were annual meetings on quantitative biology. The early symposia went on for weeks, sometimes just one lecture a day with copious discussions. Robert Furchgott (M98)[27] attended such a symposium in 1938 for the first time. It was on protein chemistry, with lots of famous scientists around such as Edsall, Fruton, Astbury, Mirsky, Langmuir (C32), du Vigneaud (C55), and others. Furchgott tended the lantern-slide projector, for which he was given free room and board. Once there, he was free to take part in the discussions and Furchgott even voiced his disagreement with some of Langmuir's conclusions about protein structures deduced from his monolayer studies. Furchgott's doctoral project grew out of his experience at Cold Spring Harbor. In 1939, the symposium was on hormones and metabolism, with the speakers including Leonor Michaelis, Fritz Lipmann (M53), Carl Cori (M47), and others. By 1940, Furchgott had already completed his doctorate at Northwestern University and was one of the invited speakers. His interactions at the symposium landed him a postdoctoral position at Cornell Medical School in New York City.

Many years later, a Cold Spring Harbor symposium gave François Jacob (M65) the first taste of what it meant to have a relaxed atmosphere between professors and students, when he was invited to the meeting from Paris.[28] This was May 1953, and Jacob heard for the first time about the double helix model of the nucleic acid structure. Jacob sees the rigid atmosphere in French academia, especially in the universities, as one of the main obstacles to more French Nobel Prizes in the post-Second World War period. The Pasteur Institute, Jacob's base, is less formal and rigid than the universities and it has a better record in discoveries, and Nobel Prizes as well.

Rita Levi-Montalcini (M86) also noted the difference in atmosphere between American and European laboratories. After having spent years at Washington University in St Louis, she returned to Italy: 'Accustomed as I had become to the cordial "Hi, Doc" of American technicians and students, I was embarrassed by the ceremony which, at the beginning of the 1960s, still regulated relationships between professors and the rank and file [in Italy].'[29]

By design

In the last thirty years the United States' National Institutes of Health (NIH) has become one of the world's foremost research institutions in biomedical sciences. During the period immediately after the Second World War there were already several future Nobel laureates working at what later became the NIH, although it was a much less active and a much less recognized institution at that time. Arthur Kornberg (M59) spent years in one of NIH's institutes doing nutrition research and found his work increasingly uninspiring. When he attended a seminar about genes by Edward Tatum (M58) it was a revelation for him. The senior scientists around Kornberg did not think the new biochemistry important so Kornberg and three other young scientists, L. Heppel, B. Horecker, and H. Tabor, formed an informal circle. They met every noon for lunch and reviewed the current literature very seriously; they taught themselves modern biochemistry. They read and examined the *Journal of Biological Chemistry* from cover to cover.

One of the decisive discoveries in molecular biology (mentioned before, p. 98) was made in the NIH before it became a leading research institution. Marshall Nirenberg (M68) and Heinrich Matthaei took the first step in cracking the genetic code in 1961. They were knowledgeable and lucky and the NIH proved to be a most constructive environment.

The sample of poly U they used was a gift from a colleague there, D. Bradley. L. Heppel proved to be a great help; he gave Nirenberg the first triplets and the idea of an enzymatic method to synthesize all 64 triplets. Nirenberg understood immediately the tremendous importance of their cracking the genetic code, but he also knew that he had to be very convincing in presenting it to the scientific community. If the discovery itself was serendipitous, proving the identity of the protein produced was not lacking serendipity either. Once again, the NIH proved to be an ideal place for that. Nirenberg wanted to find out something about the physical properties of polyphenylalanine. On his way to the library he went down to the floor below his lab, where Christian Anfinsen (C72) had his lab. Anfinsen was to win the Nobel Prize for his work on the structure of proteins. Nirenberg did not find him in the lab, but he found a young visitor, Michael Sela, who was spending his sabbatical with Anfinsen. He later became president of the Weizmann Institute in Israel. Sela was very knowledgeable about synthetic peptides and he used them in immunology. He told Nirenberg that polyphenylalanine was very insoluble, that it wasn't soluble in normal solvents but did dissolve in 15% hydrobromic acid dissolved in glacial acetic acid. Sela just happened to have some of this solvent handy and offered it to Nirenberg.

It was only about 20 years later that Nirenberg found out the full story. Polyphenylalanine was a reagent used in the study of proteins, and Sela, in the process of carrying out an assay, had made a mistake. He poured a suspension of polyphenylalanine into the wrong solution, the particular combination he had described to Nirenberg, and the polyphenylalanine dissolved, to Sela's surprise. As it turned out, on that particular day, on his way to the library, Nirenberg asked the only person in the world who could have answered his question. To this day he knows of no other solvent for polyphenylalanine.

Another fortunate feature of Nirenberg's situation was that the NIH supported his research and he could do whatever he wanted. He did not have to write grant applications, and had he applied for one, chances are that he would have been refused on the grounds of completely lacking experience in his proposed research project. When he applied to attend a Cold Spring Harbor symposium, he was turned down because he was unknown in the field and the NIH was not yet a highly recognized research venue.

The Rockefeller University, and its predecessor the Rockefeller Institute, has been exceptionally strong in biomedical research and now boasts over 20 Nobel laureates. The latest addition was Günter Blobel in 1999 in physiology or medicine. The institute was organized in 1901 and its hospital started a little later, in 1910. The first director of the hospital was Rufus Cole, who selected people like Oswald Avery. The time was ripe for building up a strong institution.

It all started with John D. Rockefeller, Senior, who originally accumulated the family fortune. He had an adviser, Frederick Taylor Gates, a Baptist minister who had read a book on medicine by Sir William Osler in the 1890s. Osler, a Canadian who came down to the United States and was involved with the building of the Johns Hopkins Medical School, had written a famous textbook on medicine. Having read the book, Gates realized the primitive state of affairs in medicine: in so many cases it was only possible to describe the diseases, and nothing could be done about treating them. Gates persuaded Rockefeller to set up an institute for medical research.

Support for research for the first 50 years came totally from endowment. The federal government started supporting biomedical research with the National Institute, later, Institutes, of Health after the Second World War. At the Rockefeller Institute they started thinking about external support in the mid-1950s. Researchers used to be able to concentrate on their projects for long periods without worrying about funding. A typical example was Bruce Merrifield (C84); when he came upon the idea for solid state peptide synthesis, he was free to work for years on it, and on the method that led from it, which brought him the Nobel Prize. He did not have to worry about funding or about working on a project for three years without a publication.

The solid state peptide synthesis is also an example of long lines of related research at the Rockefeller.[30] Max Bergmann was the foremost peptide chemist in the world for several years. He was a student in Germany of Emil Fischer, who was the father of the field, and came to America just before the Second World War. He was recognized to be important and the director of the institute offered him a position. Bergmann assembled a strong group, including Stanford Moore (C72) and William Stein (C72) who devised an amino acid analysis method and built an automated machine.

An interesting titbit about the Rockefeller Institute is that it paid a lot of attention to how people wrote their reports. John Maddox,[31] later the editor of *Nature*, worked at the institute for a year in 1962–63 as a member of the academic staff. His task was to advise people on how to write scientific papers, and he persuaded many who had written poor papers not to publish them. Maddox met a series of important scientists during his year at the Rockefeller, who either worked there or had come for a visit. They included George Uhlenbeck and Samuel Goudsmit, the discoverers of the electron spin, F. Peyton Rous (M66), the discoverer of the sarcoma virus, Alfred Mirsky, a former vocal opponent of Oswald Avery's findings that DNA was the hereditary material, and David Baltimore (M75), who just joined the Rockefeller as a PhD student.

Personal imprints

There are famous and less famous venues; some are in the forefront of world science for a sustained period, while others come up from time to time in various subjects. The College of Chemistry at the University of California, Berkeley was one of the latter. It was originally G. N. Lewis who developed it into a strong setting for chemistry. Henry Taube (C83) remembers:

> Just the general atmosphere of the college was conducive of this; chemistry was in the air. When members of the faculty would meet, let's say in the halls, the discussion would be about some seminar topic that hadn't been quite resolved or about a puzzling observation that someone had made, sometimes in the freshman laboratory. What got to me early was not only their enthusiasm for the subject but their frankness in admitting the limits of their understanding and their willingness to learn from each other. There was little pretense and they didn't feel that they had to impress others. At any rate, the fire was lit there quite early in my stay.[32]

Donald Cram (C87) takes a critical view of another great venue, the graduate school at Harvard, which in the period just after the Second World War had not yet developed into what it later became:

> I was better prepared for Harvard than Harvard was prepared for me. Some of their postwar staff was missing and they were just getting organized. My first two days there I took four qualifying

examinations and passed them, which allowed me to plunge into
research and a minimum of advanced, stimulating courses. Bob
Woodward and Paul Bartlett's courses were particularly exciting.
My year at Nebraska and three years at Merck made Harvard
easy.[33]

When E. J. Corey (C90)[34] joined Harvard's chemistry department in
1959, the faculty included Paul D. Bartlett, Konrad Bloch, Louis Fieser,
George Kistiakowski, E. G. Rochow, Frank Westheimer, E. B. Wilson,
and R. B. Woodward. Woodward's research seminars were legendary, and
he (C65) and Bloch (M64) were soon to be Nobel laureates.

George Olah's (C94) career benefited from existing good venues as
well as from those he was building himself. He was Géza Zemplén's
student in Budapest and Zemplén had studied under Emil Fischer (C02)
in Germany. Olah participated in organizing what was to become the principal chemistry research venue of the Hungarian Academy of Sciences,
and took a leading position in it. When he moved to Cleveland and became
chairman of the chemistry department at Western Reserve University,
he initiated its joining with the chemistry department of the Case Institute
of Technology. Finally, at the University of Southern California, which
is not the foremost academic institution in the region, he built up a
basic research institute for hydrocarbon research, not entirely unlike the
Budapest institute of three decades before. He likes to point out that what
is now Cal Tech used to be a provincial engineering school, and that it
only took a few individuals and a few decades to convert it into a world-
renowned research institution.[35] Today, southern California has one of
the world's greatest concentrations of strong venues. As well as Cal
Tech there is the University of California at Los Angeles and further
south the Scripps Institute and the Salk Institute, plus other campuses of
the University of California. Although there is this extraordinary gathering
of great science in southern California, Paul Boyer does not find it to be
an intellectual center in the sense that great scientists get together in any
organized way on a routine basis. The interactions are compartmentalized,
according to one's profession.

Other venues have also been very important in nurturing talent and
Nobel laureates. Alfred Gilman (M94)[36] went to Cleveland after his undergraduate degree from Yale. He was interested in medicine but wanted to
do research. It was common in those days to take a medical degree and

then get training in research. By fortunate coincidence, a friend of Gilman's father, Earl Sutherland (M71), had established a seven-year combined MD/PhD program at Case Western Reserve University (by then the two schools had joined). It was a small group and Sutherland recruited every one of the students personally. Gilman benefited greatly from this special program, although Sutherland left Cleveland soon after he had established it.

Paul Boyer (C97)[37] did his graduate studies at the University of Wisconsin. He thinks that in his time there Wisconsin had a much better biochemistry department than most other universities. The University of Wisconsin appears on the career records of quite a few Nobel laureates, even if, for most, only temporarily, including Eugene Wigner (P63), Joshua Lederberg (M58), Gobind Khorana (M68), Howard Temin (M75), Michael Smith (C93), John Van Vleck (P77), John Bardeen (P56, P72), P. Debye (C36), and others, as well as Boyer. It seems to provide a stimulating environment but fails to hold on to its laureates in the long run.

Not only academic laboratories could be good venues conducive to Nobel work. John Vane's (M82) prize-winning discoveries of prostacyclin and of the aspirin/prostaglandin interaction were made in the laboratories of the Wellcome Foundation in England, and so were some of James Black's (M88) discoveries. Gertrude Elion (M88) and George Hitchings (M88) spent their entire career in the Wellcome laboratories of the Burroughs Wellcome Company in the United States. An earlier research director at a Wellcome laboratory, Henry Dale (M36), also won the Nobel Prize. After his experience at the Wellcome laboratories he moved on to a brilliant career, which included the presidency of the Royal Society.

European experience

Moving around most of Europe in search of the best and varied venues involves changing countries, and often languages. Odd Hassel (C69)[38] entered the University of Oslo in 1915 and majored in chemistry in 1920, with minors in math, physics, and mechanics. He then spent some time in France, Italy, and Germany. Hassel attended the physics lectures of Paul Langevin in Paris, worked with Kasimir Fajans in Munich, and did his doctoral work in X-ray crystallography at the Kaiser Wilhelm Institute in Berlin-Dahlem. For his doctorate, Max von Laue (P14) examined him in physics and Fritz Haber (C18) in chemistry. He returned to Norway in 1925.

Nikolai Semenov's (C56) early years in research were unique. This was the time of revolutions and civil war in Russia, and many thought that under the circumstances scientific research would be suspended. But the new Soviet power recognized its importance, created the minimal conditions for independent work, and secured the necessary funds for international monographs and journals. This is how Semenov remembered the early 1920s in 1965:

> This was a heroic era indeed. It was moving and uplifting to see the thirst for science of our impoverished, tortured, and liberated people. We received letters from the most remote corners of the country. If somebody read or created something of interest, he let us know at once. The scientific institutes themselves turned often to the public for recruiting new coworkers. Although we could hardly pay them, people were joining us *en masse*. . . . Everybody was willing and ready, even if living merely on bread and water, to participate in creating the new science. It is also true, though, that the young researchers were given greater independence than is customary today.[39]

Just as in France the Pasteur Institute appears to be more conducive to Nobel-level research than the universities, so the Max Planck institutes in Germany have this advantage over the universities. Manfred Eigen (C67)[40] did his Nobel Prize-winning research in post-war Germany, and by the mid-1960s there was the opportunity for a great renewal of German science. The Max Planck Society was building new institutes, which then became the hotbeds for most of the German Nobel Prizes in later years. Eigen contributed to creating the new venues in Göttingen. This was at the time when he was receiving attractive offers to go to the United States, but people tried to keep him in Germany. He argued that biochemistry and biology were too weak in Göttingen so he was charged with founding a new institute; Eigen wanted this to be close to the university, which had just moved its campus out to the countryside. They united three groups in the new institute, physical chemistry, spectroscopy, and neurophysiology. For a unifying subject, they chose the study of the central nervous system at both the molecular and the cellular level. The Institute for Biophysical Chemistry opened in 1970, and its success has been demonstrated by Fitzpeter Schäfer's invention of the dyelaser, and Jens Frahm's application of nuclear magnetic resonance tomography to examinations of the whole body. Erwin Neher (M91) and Bert Sakmann (M91) made important discoveries concerning the single ion channels in cells. Eigen does research both in California and Göttingen, and he makes an interesting comparison between the two. A piece of work done in the United States is pushed more than when it is done in Germany. There is a much higher concentration of science now in America than in Europe. Göttingen had its best years some time ago, and Hitler contributed to diminishing its importance.

Rudolf Mössbauer (P61)[41] believes that there is a political reason why research in Germany is stronger in the institutes of the Max Planck Society than in the universities. He complains that the German laws introduced in 1976, in the aftermath of the student unrest around 1968, bind the universities in various ways. For example, the university professors are civil servants and the same rigid rules apply to them as to the people who work in the post office or finance office. Although universities need more liberty and flexibility, Mössbauer admits that the majority of professors would vote for the present system because they have got used to it. A full professor may do nothing for the rest of his life and nobody will bother him. Competition is reduced to zero. The Max Planck institutes are better off in having more financial assistance and greater liberty. The superiority of the Max Planck institutes versus universities in research does not necessarily translate into higher productivity. It has been demonstrated that, during a certain period, a citation of research produced by the Max Planck Institute for Biochemistry in Martinsried was twice as expensive as that produced by the Institute of Genetics at Cologne University.[42] The Max Planck Society gets 45% of their income from the federal government, 45% from the states, and 10% from industry. The 10% gives them the flexibility. Mössbauer prefers the American system, which relies primarily on the universities for basic research. When you are young, you are more active in research and when you get older you might become more active in education. These two areas are in harmony in the university. He believes that the most important difference between the American and German systems is the lack of private universities in Germany. Cal Tech, where Mössbauer spent years, is a good example of the private universities that are setting the standards for the state-operated ones. Berkeley, for instance, is a state university, part of the University of California system, and others include UCLA and Irvine. These state campuses are at such a very high level because they must compete in California with the likes of Stanford and Cal Tech.

Mössbauer finds, though, many strong features in European science, where often several countries work together. For instance, Italy and France are especially strong in high-energy physics and the CERN Laboratory in Geneva is outstanding. There are several reasons for its excellence: it has a unique arsenal of instruments, and it is operated as a user's institute, where the experimental proposals originate entirely from within its scientific community. It is, then, a special European advantage that the Germans, French, and British represent very different mentalities. Working together, they can do miracles. The countries may be small compared with America, but what matters is that they think differently.

Right conditions

Great venues for scientific research often include great support personnel. A support person may be another Nobel-caliber scientist, a lab steward, a skilled workshop mechanic, a secretary who liberates the scientist from a lot of chores or a department chair who does the same, and so on. Linus Pauling had the supporting activity of Robert Corey, one of America's best-trained X-ray crystallographers. According to Richard Marsh[43] of Cal Tech, a former graduate student of Pauling and himself a noted X-ray crystallographer, Pauling and Corey had a close scientific association but were distinctly opposite in personality. Pauling had a wonderful chemical knowledge, instinct, and intuition. He could construct a coherent and logical manuscript during a single dictation. Corey always wanted to be absolutely certain, fussing with every word in a manuscript and adding several caveats. However, Pauling understood his own impetuosity and realized the need to restrain it. If Corey's nit-picking frustrated him, he never showed it. Pauling was flamboyant, Corey was quiet, Pauling liked the limelight, Corey liked to work in the quiet of the laboratory. Corey provided the data, Pauling the interpretation. They never interacted socially and their wives were just as opposite in personality and interest as their husbands. Although Pauling called Corey 'Bob', Marsh never heard Corey use anything other than 'Dr. Pauling'. Corey first came to Pasadena to work with Pauling for a year, initially supporting himself with the money he had received when he lost his job at the Rockefeller Institute. Then he got a position at Cal Tech and continued his X-ray crystallographic studies throughout his career. Corey was quiet, hard-working, uninterested in outside affairs. He and Pauling complemented each other, and Corey provided a solid and reliable background for Pauling.[44] The Cal Tech Archives stores very few documents from Corey because he did not preserve his papers. There is, though, a nice get-well note from Pauling, dated 26 February 1968. The note is typed, but then on 5 March a PS is added in longhand: 'I appreciate very much your fine contribution to Structural Chemistry and Molecular Biology.' Signed, Linus.[45]

Family members often make great support personnel, whether as fellow scientists or just as 'home-makers'. Speaking about the support provided by spouses may sound politically correct, but it is almost exclusively wives who support husbands and not the other way around. At home Dorothy Hodgkin (C64) was wife and mother before all else, and everything revolved around her husband. Only her Nobel Prize made their children realize that she was also a distinguished person to the rest of the world.[46]

Arthur Kornberg's (M59) wife, Sylvy, was also his closest co-worker. She had contributed significantly to the science surrounding the discovery of DNA polymerase. She was quoted as having said, upon hearing about her husband's Nobel Prize, 'I was robbed.'[47]

For Jerome Karle, (C85) his wife and fellow crystallographer, Isabella, has been a great support, and she could have been a co-winner of the Nobel Prize in Chemistry for 1985. Isabella was a principal player in applying the technique of direct methods in X-ray crystallography for which Herbert Hauptman and Jerome Karle were awarded the chemistry prize in 1985. Jerome stated: 'During my entire married life I have had the strong support of my wife, both technical and spiritual.'[48]

John Cornforth (C75)[49] was also blessed with most beautiful support from his wife, Rita, on several levels. Both of them came to England from Australia, each on an 1851 Exhibition Scholarship, and both did their PhD studies with Robert Robinson (C47). Their joint work led to many shared successes, although only John shared the Nobel Prize in Chemistry with Vladimir Prelog in 1975. Rita Cornforth has also acted as her deaf husband's interpreter.

George Olah's (C94)[35] wife, Judith, has been a solid support. Originally she was not trained as a research chemist but George enrolled her at the university, preparing her for a more important role. Herbert Brown (C79)[50] and his wife Sarah provide another model of a fine husband–wife support team. First they were fellow students, but then Sarah took a job to support Herbert in his further studies. She has been involved in all facets of her husband's professional life during their marriage of over six decades, including advising on, financing, and protecting his work and interests.[51]

Earlier in this chapter we saw that Gisela Perutz provided support for her husband Max by running the LMB canteen, which proved to be an important place for interactions among the LMB staff. Max Perutz seems to have had an unusually large number of support personnel throughout his career. People suggested to him the topic for his doctoral work, others gave him the necessary sample, while still others contributed crucial ideas to solving the problems of protein crystallography.[52] While Perutz was always extremely hard-working and relentless in his quest for uncovering the structure of hemoglobin, he remembers and publicizes his debts to others more than is customary.

William Lawrence Bragg was 25 when he became co-recipient of the Nobel Prize in Physics for 1915. He shared the prize with his father, William Henry. Many have thought that the son's fortune was linked to that of his father's, but in reality the junior Bragg was the pioneer and his father went along. William Henry was professor of physics in Leeds, and had constructed an X-ray spectrometer with which the two Braggs then solved the structure of diamond and other simple crystalline substances. By then W. L. Bragg had disproved his father's corpuscular theory of X-rays, discovered the Bragg equation, interpreted the diffraction patterns obtained by Laue's co-workers in Munich, and solved the structure of the sodium chloride crystal. It was more a case of the father being the supporting personnel for the son. This reverse kind of interaction may have introduced some strain into their relationship, which the younger Bragg never alluded to until shortly before his death. At that point he wrote to Perutz, his former pupil, 'I hope that there are many things your son is tremendously good at which you can't do at all, because that is the best foundation for a father–son relationship.'[53]

It is hard to believe that the first full publication by Charles Pedersen (C87), reporting on his syntheses of the new crown ethers, was the work of a sole scientist rather than of a team. He spent his entire 42-year career at Du Pont, where his support personnel consisted of a sole technician, who was able though technically untrained. Curiously, his managers might also be counted as his support personnel, since they allowed him to carry on his fundamental research, although company policy, alas, delayed its publication by a few years. When Pedersen was awarded the Nobel Prize at the age of 83, he was too frail to travel by ordinary means but Du Pont flew him to Stockholm in a private plane.[54]

The right environment may be synergistically beneficial for individuals. For example, Washington University in St Louis, Missouri, was a great venue for many scientists in biomedical fields. It had several laboratories, in the best known of which were the Coris, Carl and Gerty (M47). In another was Viktor Hamburger, who stimulated his associates to perform more than they would have thought themselves capable of. There was a tribute to the venue created by Hamburger in what Stanley Cohen (M86) told Rita Levi-Montalcini (M86): 'You and I are good, but together we are wonderful.'[55] Watson and Crick have been compared to two elements that were mildly radioactive in isolation but which, when brought together in the Cavendish Laboratory, began to implode.[56]

8

Mentor

One of the most common features about the research careers of Nobel laureates is that many have worked at some point under a former or future Nobel laureate or with someone closely associated with a laureate. Hans Krebs (M53) constructed a genealogy of Nobel laureate biochemists and all, without exception, came from laboratories of Nobel laureates or future Nobel laureates.[1] When John Walker (C97)[2] finished his chemistry studies, he became a PhD student of Edward Abraham in Oxford. Abraham had been part of the penicillin team during the Second World War, and a PhD student of Robert Robinson (C47), before he went to work with Boris Chain (M45). When Walker visited the Justus von Liebig Museum in Giessen, Germany, he realized that through Abraham and Robinson he descended from the great German chemists in a scientific genealogy. Walker appreciates how Abraham guided him in reading appropriate literature, like the paper by Jacob (M65) and Monod (M65). Such guidance was crucial as there were no formal requirements or mandatory course work; it was only the opportunity that was provided. Edward M. Purcell (P52) is another good example of a Nobel lineage member. John Van Vleck (P77), who was one of his teachers in graduate school, had been involved in developing microwave radar during the Second World War, heading a group of researchers in which he succeeded Norman Ramsey (P89). The division head to whom Purcell reported was I. I. Rabi (P44). Later, at Harvard, Nicolaas Bloembergen (P81) was one of Purcell's graduate students.[3]

The kind of genealogy demonstrated by Krebs, Walker, and Purcell is remarkable, if hardly surprising. It might be argued that the topics chosen by a former or future laureate proved to be very important. It might also be argued that the topics chosen by a Nobel laureate were *assumed* to be very important, or even that the laureate promoted his protégé in the most successful way. Even when the laureate is no longer in his prime for research, his name and prior achievements bring in excellent graduate students and postdocs and often, though not always, ensure the necessary research support. The job prospects of Nobel laureates' students are enhanced. One of Daniel Nathans' (M78) graduating students was given a job immediately after Nathans' Nobel Prize, a job for which the student had been rejected a couple of weeks previously.

By example

The best mentors influence their pupils by example, not instruction, and they encourage their independence. One exception in this is the preparation of manuscripts, which is a crucial step in scientific research as it constitutes the main link between the laboratory and the outside world. The good mentor ruthlessly revises his budding disciples' first writings. He is also fair, even generous, in assigning credit and letting his pupils shine. Bruce Merrifield (C84) was allowed to publish his invention alone, bringing him instant recognition. The same was true of Paul Berg (C80) and Joshua Lederberg (M58). Gertrude Elion (M88) did not put her name on the papers of her young associates merely because they worked for her. She became a co-author only if she had truly participated in the work.

James Watson (M62) and Walter Gilbert (C80) rigorously adhered to the principle that their names would not appear on the papers of their students unless they had manually taken part in the work. Gilbert learned this from Watson, who has strong opinion about this, saying that people perform best when they are working for themselves. However, it is not easy to follow such a principle and it demands considerable self-restraint. This approach somewhat underestimates the intellectual contribution and over-emphasizes the laboratory work. Nonetheless, it is attractive, especially in the light of the more common exploitation of graduate students by their supervisors. One of Gilbert's former associates tried for years to follow the practice back in his own European laboratory, only to give it up eventually for practical considerations.[4]

Watson himself emulated what he had seen in Salvador Luria's (M69) attitude while Watson was a graduate student at Indiana University. Luria, as well as Renato Dulbecco (M75), had come to Indiana from the University of Turin, Italy, where Giuseppe Levi was their mentor; Levi had also served as mentor for Rita Levi-Montalcini (M86). Guiseppe

Levi's presence and his willing students made Turin another exceptional venue, if only for a short time.[5] Levi-Montalcini shared the Nobel Prize with Stanley Cohen (M86), who learned isotope methodology from Martin Kamen, the co-discoverer of carbon-14 (see p. 231), at Washington University. Cohen also benefited from participation in the journal club organized by Arthur Kornberg (M59) at Washington University. Then he joined Viktor Hamburger and Rita Levi-Montalcini (M86). Hamburger also served as mentor at Washington University.[6] He might have been the 'third person' in the 1986 Nobel Prize in Physiology or Medicine shared by Cohen and Levi-Montalcini. Luria and Dulbecco had a strong impact on Susumu Tonegawa's (M87) career. Tonegawa did postdoctoral work with Renato Dulbecco (M75) at the Salk Institute in La Jolla, and then with Niels Jerne (M84) in Basel, where Dulbecco had sent him. Then Luria (M69) was instrumental in bringing Tonegawa to the Massachusetts Institute of Technology.[7]

Frederick Sanger (C58, C80) did not have a university professor's career; he spent his life as a researcher in Cambridge, and acted as an adviser for a few research students. One of his colleagues, César Milstein (M84), described him in the following way:

> In more modern times, I consider an extraordinary scientist, by all standards, Fred Sanger, in a very strange way. He is a very quiet person but he did and initiated a tremendous amount of work, and his impact in science has been enormous. He had one simple thing in mind that you have to break the limits of technological capabilities, to work out methods, to do new things. This was his philosophy. He was an absolute pioneer. When he first sequenced a protein, it was not yet clear whether proteins were made by individual amino acids or not. When he did the N-terminal sequences it was not known that you could actually do such a thing. Once it became clear that the amino acids in proteins were in a certain order, the whole problem of the genetic code became essential. But Sanger was not interested in finding out that, his interest was in breaking the boundaries. He wanted to show that you could do sequences, if they exist. He then went on, in the same way, for methods of doing RNA and DNA. To me he is the most spectacular experimental scientist. Although he is a quiet and simple person, he was a very inspiring teacher. When you talk to him, you may not think so, but he had a tremendous insight into problems.[8]

Coming in all varieties

The teacher/pupil relationship is a two-sided interaction. It not only takes a good teacher but also a willing student to make it work and be productive. A teacher may be the graduate adviser, the first boss, or anybody else who is worth observing and emulating. The source of education need not even be a person in proximity. Thus, for example, Vladimir Prelog (C75)[9] considered Robert Robinson (C47) as his mentor because Robinson's papers and his way of doing natural compounds chemistry and reaction mechanisms very much influenced him. For a long time Robinson's image decorated Prelog's office. Then, shortly before the end of his long life, Prelog replaced Robinson's photo by one of Leopold Ružička (C39), Prelog's teacher and predecessor at the Federal Institute of Technology (ETH)[10] Zurich. Prelog worshipped Robinson in a way he never worshipped Ružička, but Prelog pointed out that for worshipping you need some distance and Ružička was a little too near to him. When James Watson (M62) was 23 years old, he went through a phase of wanting to understand how Linus Pauling thought. He wanted to write papers in Pauling's style. He thinks that imitation is important. 'Even though imitation might initially sound the cheap way to go, it is only a question of whom you imitate. If you are trying to mimic someone very good, you really can't pull it off, but what comes out might nevertheless be fairly interesting.'[11]

The single most important aspect of a scientific career is one's teacher, taken in the broadest sense. The most common case is when the mentor is the graduate adviser, who is called by different names in different situations, like supervisor, patron in French, and Doktorvater in German. A book, a casual remark, an idol, or, sadly, even a negative example, may play the mentor's role. Such adverse examples figured in Donald Cram's (C87) early career and he remembers them without bitterness.[12] Aaron Klug (C82)[13] was already a PhD when he worked with Rosalind Franklin in Birkbeck College, London. She introduced him to the study of viruses and was his mentor, but she could not be his supervisor officially because she did not have a university appointment. His official supervisor, J. Desmond Bernal, was too often away and otherwise preoccupied to act as an ideal mentor. The combination of the two was, however, most fortunate. Franklin was meticulous and demanding, both of herself and others. Bernal's frequent absences were amply compensated for by the attraction his personality had for the most interesting visitors to his college, people, like Linus Pauling, Pablo Picasso, and Buckminster Fuller. In a very indirect sense Fuller also served as Klug's mentor in his virus studies. When Caspar and Klug found the icosahedral virus structure, they stated, 'The solution we have found was, in fact, inspired by the geometrical principles applied by Buckminster Fuller in the construction of geodesic domes.'[14] Klug was a willing pupil. When others turned away from Fuller's endless and chaotic lectures at Birkbeck College, Klug persevered and worked hard to understand him.

J. D. Bernal was one of William Henry Bragg's (P15) research students, and Bragg may have had Bernal in mind when he wrote: 'A good research student is like a fire, which needs but a match to start it.'[15] Several of Bernal's disciples became Nobel laureates. Bragg's observation as mentor is mirrored by a pupil, Jean-Marie Lehn (C87), who said of his chemistry professor, Guy Ourisson, that he provided 'the spark that lighted everything up'.[16]

Arthur Schawlow (P81) spent two years in the physics department of Columbia University under I. I. Rabi's (P44) leadership. Eight Nobel laureates-to-be worked there during those two years.[17] Rabi became a great mentor almost by default.[18] He was away from Columbia University during the war years, and when he arrived back from military service the once great physics department had almost disappeared. Rabi became chairman and wanted to recruit senior scientists, but they had already taken appointments elsewhere. That is when he decided 'to begin by developing our own young people and bringing in other young people'.

Ahmed Zewail (C99), who did his undergraduate studies in Alexandria, Egypt and went to graduate school at the University of Pennsylvania, has only praise for his teachers. As his most important mentor he singles out a later influence, Richard Bernstein, whom he came across when Zewail was already a well-established professor at Cal Tech:

> Dick was a first-rate experimentalist, and one of the founders of
> molecular-beam scattering. He had high integrity, worked all the
> time, had a commitment to science and family, and was especially
> helpful to younger colleagues.
>
> My relationship with him was my good fortune. I was a
> newcomer in molecular reaction dynamics when he happened to be

visiting Cal Tech. First he was a colleague, then a supporter, and, ultimately, my friend. He took two sabbaticals with me. He was in the 'tenth excited state' about femtochemistry. He always believed in the 'unlimited opportunities' of this field, and these are his words. At that time, hearing this from a well-established and well-respected senior was a tremendous boost for me. You hear about how difficult it is sometimes to make a breakthrough in other fields, especially if you are an outsider. You may be seriously hindered by hostile attitude or simply by being ignored. Dick Bernstein saw to it from day one that this was not the case, and he made prophetic predictions about the field. During his first sabbatical, we did collaborative work here. Every day when he was in town and I was in town, there would be a pot of coffee and we would sit in my office and talk. It was a fantastic experience talking with somebody of some 30 years of experience and seeing that I was stimulating him. During his second sabbatical, just before he died, we built a new apparatus and wanted to study the femtochemistry of surfaces, and we had all kinds of plans to work together. What can I say? He is a hero.[19]

Bernstein was apparently someone who had an impact on people he came across. Robert Whetten, his junior by 35 years, had many conversations with him during the six years they spent together at UCLA. For Whetten these encounters were never comfortable or relaxing.

> He had the attitude that life is short and you shouldn't waste it with frivolous things and that people who had some kind of talent should be working basically all the time to reach the highest level . . . . Bernstein liked to say, you'd be recognized in the end for one thing, maximum one thing . . . so the point is to find this and stop wasting your time on all other things.[20]

For years after his death, Whetten kept hearing Bernstein's irritating voice in the back of his head, measuring up his deeds against what Bernstein would have expected him to do.[20]

The Italian–American Enrico Fermi (P38) was a great mentor to many in physics. Emilio Segrè (P59)[21] was one of his famous pupils and their teacher–pupil relationship originated back in Italy. Segrè entered the University of Rome as an engineering student in 1922. He knew modern physics and the physics in Rome did not appeal to him at all: as the instructions became more professional, they also became less interesting. He calls it 'a tremendous stroke of luck, something like winning the Irish

Sweepstakes', that Fermi came to Rome in 1927 looking for students. When he offered to teach Segrè physics, Segrè realized at once that this was a unique opportunity. Fermi had a small group of students, which also included Amaldi, Rasetti, and Majorana, and gave them private lessons two or three times a week. Fermi continued this practice for years, repeating it later at Columbia, and later yet at Chicago. He was rather ruthless in choosing his students: he was prepared to devote an infinite amount of effort and pain to them and did not want his efforts wasted. After graduation Fermi suggested research topics to Segrè but also encouraged him to find his own direction. Segrè started doing spectroscopic work; for some of the experiments he went to Pieter Zeeman (P02) in Holland, and then to Otto Stern (P43) in Hamburg, before returning to Rome.

In October 1934 Segrè participated in the discovery of slow neutrons in Fermi's laboratory at the University of Rome. It was a discovery in one day. In the morning they did not know about slow neutrons, then Fermi understood what happened, and in the evening the paper was written and sent. The observation was that when they put a piece of paraffin between a radioactive source and an object, there was a great increase of radioactivity in the object. Fermi and his students went to lunch and took their siesta and when they came back Fermi said, 'I know.' The neutrons are slowed down by collisions with hydrogen atoms in the paraffin. The slow neutrons, with their random motion, have more time for, and a higher probability of, collisions in the object to generate radioactivity than fast neutrons. The next day they filed for a patent and Segrè went around Rome with a basket, visiting chemical laboratories, collecting all the elements to try them systematically for their ability to slow down the neutrons.

Segrè went to Berkeley in 1938, and Fermi to Columbia in early 1939. In late 1940 Segrè visited Fermi, who told him about his hunch that element 94 might be a slow neutron fissioner. Segrè then worked to prepare a sample, of what was to become known as plutonium, to see if it was fissionable or not. By the spring of 1941 they had enough to make a thin layer; this was sufficient to do some chemistry on it, and they found that it was fissionable, similar to uranium-235. That opened new horizons for another nuclear fuel. In the beginning the atomic business in the United States was left to foreigners, people like Wigner (P63), Fermi (P38), Segrè (P59), Teller, Bethe (P67), and others. Then it became necessary to build up a big organization, gradually leading to the Manhattan Project in 1942. By the summer of 1942 Segrè had his own co-workers, and they started making the physical measurements necessary for building a bomb.

Owen Chamberlain[22] was Fermi's scientific child and grandchild in one. He was doing graduate studies in Berkeley, but then the war interrupted and he joined the Manhattan Project. E. O. Lawrence (P39) assigned him to be a helper to Emilio Segrè. Chamberlain found this more valuable than going to graduate school. Their job was to test some of the conditions that might affect the final bomb design, such as the spontaneous fission rate of nuclei. This work was part of the Manhattan Project being carried out at Berkeley. In the middle of 1943 Chamberlain moved, with Robert Oppenheimer, Hans Bethe, and many others, to Los Alamos, where he had a lot of contact with both Segrè and Fermi. Once Chamberlain shared a long train ride with Fermi, and for a day and a half he had Fermi at his disposal, answering all his questions about physics.

When the war was over Chamberlain stayed at Los Alamos for another six months, during which time they set up the Los Alamos University. Fermi gave a course on electromagnetic theory and Bethe was also among the teachers. Chamberlain found some of the courses too difficult and he realized that Bethe was aiming his lectures at Fermi, rather than at his students. Edward Teller was also at the University of Chicago when Chamberlain was there. In the morning Teller would describe the work he planned for the hydrogen bomb, and in the afternoon Fermi would lecture on it, trying to demonstrate that it would not work. When they completed their operations in Los Alamos, Fermi moved back to the University of Chicago and Chamberlain joined him as a graduate student. When he finished, he returned to Berkeley and to Segrè's group. Ernest Lawrence (P39), too, was a mentor to Segrè and Chamberlain, as he was to many others, including Edwin M. McMillan (C51). McMillan, whose first scientific publication was made at the suggestion of Linus Pauling (C54), spent most of his career in close association with Lawrence. He had extensive correspondence with Bethe in the 1930s, which involved a large number of calculations, and at some point in the same decade, McMillan enlisted Segrè into his experimental studies in nuclear chemistry. When McMillan was called away for military duty, Glenn Seaborg (C51) continued his work on the discovery of plutonium; they published the work together and shared the Nobel Prize.[23]

Lucky pupils

Further examples illustrate the fate of some fortunate students who found excellent mentors. John Vane (M82) went to Professor Harold Burn in Oxford. A month after Vane arrived, Burn called him into his office and said, 'Vane, I am not very happy. You must do better.' At first Vane was quite taken aback, but he soon found out that Burn did this to all his students. Vane was taught biological methods by Burn, and says of him: 'He was the father of my career.'[24]

John Cornforth (C75)[25] found Robert Robinson (C47) very stimulating. The deaf Cornforth could never learn to lip-read him because he was an extremely difficult subject, but Robinson was very good at handwriting, and would write things down for Cornforth. They started arguing from the very first day, about structures, strategies, and other science topics. As soon as Cornforth, or anybody else, put forward an idea, Robinson would immediately find an argument against it.

Francis Ryan was Joshua Lederberg's (M58)[26] mentor at Columbia College. Ryan had been a postdoctoral fellow in 1941–42 at Stanford with George Beadle (M58) and Edward Tatum (M58), their first after their initial publication in the fall of 1941 on biochemical mutants in *Neurospora*.[27] Ryan brought with him to Columbia the concept and technology of biochemical mutations. Lederberg was a sophomore in the 1942–43 academic year when he met Ryan. Deciding that he wanted to work with Ryan, Lederberg remembers that 'I gave him no option but to accept me to his laboratory. I was a real pest.' Thus Lederberg started doing research on *Neurospora* in Ryan's laboratory. Lederberg was a rather independent young researcher who could greatly benefit from his environment and professors, from the frequent communications and many seminars. Then he had become aware of the discovery, by Avery and his co-workers, that DNA was the substance of heredity, and their paper also became a 'mentor' for him.

Beadle and Tatum shared one half of the Nobel Prize 'for their discovery that genes act by regulating definite chemical events'. The other half went to Lederberg 'for his discoveries concerning gene recombination and the organization of the genetic material of bacteria'. Ryan had let Lederberg publish his results on his own and thereby gave his career a tremendous push. From Ryan's lab at Columbia, Lederberg then moved to work with Tatum, who had in the meantime moved to Yale University. As it turned out, Ryan was sandwiched between his Nobel laureate mentors and his Nobel laureate pupil. This is how Lederberg evaluated Ryan's chances for a share of the prize in hindsight:

> If he had been a more aggressive personality, he might have made sure that he was a co-author. He was not the originator of the idea of using genetic recombination but it started in his laboratory. But instead of his aggregating it to himself, in a very self-denying way he suggested that I continue the work in Tatum's laboratory. So I brought that idea to Yale with me in 1946 with Ryan's very active encouragement. I had already done the first experiments looking for recombination in Ryan's laboratory.[26]

Both Nobel laureates in physiology or medicine for 1981, David Hubel and Torsten Wiesel, came to the laboratory of the neuropathologist S. W. Kuffler at Johns Hopkins University in Baltimore, as postdoctoral fellows.[28] The degree of Kuffler's involvement can be sensed from his corrections to the first draft of Hubel and Wiesel's first abstract, which are almost more extensive than the original text. Kuffler's research was an important extension of the work by a previous Nobel laureate pair, Ragnar Granit (M67) and Haldan K. Hartline (M67), who received the prize, together with George Wald, 'for their discoveries concerning the primary physiological and chemical visual processes in the eye.' Kuffler's related study was on the receptive-field arrangements of cat retinal-ganglion cells. One might wonder whether Kuffler would have shared the prize of Hubel and Wiesel had he not died a year before their award.[29]

Roald Hoffmann (C81)[30] started out working for his PhD with Martin Gouterman at Harvard University. After about one year he went to Russia, and upon his return a year later he switched to working with William Lipscomb (C76), and finished his degree in a year, in 1962. Subsequently, during a three-year junior fellowship at Harvard, he developed his collaboration with R. B. Woodward (C65). Woodward taught him a great deal, for instance how to simplify the discussion to an absolute essential. As Hoffmann evaluates, 'It is these simple explanations that make an impact on chemists. You don't have to cloud your explanations in mathematics.' The first paper Hoffmann wrote for Lipscomb needed extensive revision, but he learned quickly and for the next paper Lipscomb had only minor changes to make. Hoffmann found writing with Woodward much more difficult. 'The language mattered to him, and he rewrote things in excruciating detail. . . . Since the collaboration with Woodward, my papers have been very pedagogic, taking the time to explain.' Woodward was not only a great mentor to his own students but also to other professors' students as well. This is the great advantage of the best venues: with a sizzling intellectual atmosphere there is cross-fertilization between the various groups. Elias J. Corey (C90) is known to be a demanding mentor at Harvard University chemistry. He did postdoctoral work on a Guggenheim fellowship in Lund in Sune Bergström's (M82) department.[31] In turn, the Swedish Bengt Samuelsson (M82), later head of the Nobel Foundation, spent a postdoctoral stint at Harvard working with Corey, and was also influenced by Konrad Bloch, (M64), Frank Westheimer, and R. B. Woodward (C65).[32]

Woodward helped Hoffmann define his own style, and George Palade (M74) did the same for Günter Blobel (M99).[33] Blobel picked up one of the questions that his mentor's discoveries had raised in his research on the structural and functional organization of the cell. Blobel's contribution was an important component in the beginning of molecular cell biology. Palade had corrected the two papers, published in 1975, for which, essentially, Blobel got the Nobel Prize. Blobel has retained the original drafts with Palade's detailed handwritten corrections and comments. His remarks were gentle: 'I would say this', 'I would not say that', 'I would do another experiment', 'Wonderful', 'Great', and so on. Palade was not a co-author.

F. Sherwood Rowland (C95)[34] went to graduate school in chemistry at the University of Chicago and came across a whole series of luminaries. His doctoral mentor was Willard F. Libby (C60). He took courses in physical chemistry from Harold Urey (C34) and Edward Teller, inorganic chemistry from Henry Taube (C83), radiochemistry from Libby, and nuclear physics from Maria Goeppert Mayer (P63) and Enrico Fermi (P38).

Manfred Eigen (C67) studied in Göttingen and Arnold Eucken and Friedrich Bonhoeffer were his idols. Heisenberg (P32) was also one of his teachers. The personal contact with his teachers was important for Eigen, as is well illustrated by the following:

> When I got the necessary accuracy, Eucken told me to take a very
> large piece of graph paper to plot everything very carefully. At

that time I was already very arrogant and I told him that my precision was much better than what he could do on paper. I liked mathematics and I went to a colleague and we developed a numerical method, which could do everything superior to plotting the data on graph paper. When I had the data and drew the line, Eucken took his ruler and checked whether my curve was properly linear or not. He said it was pretty good indeed but he spotted a millimeter difference at the end. I told him, it was his ruler and not my curve, so he went to the machine shop, had his ruler checked, and it was his ruler that was not precise enough.[35]

Two mentors had a big effect on Bruce Merrifield's (C84) life. One was Max Dunn at UCLA, who 'knew everything there was to know about amino acids', and the other was D. W. Woolley at the Rockefeller, 'a fine biochemist, a brilliant man'. Woolley had lost his eyesight in his mid-twenties due to diabetes, but carried on and maintained his lab, with a technician. He only had to hear something once, for instance if it was read to him, and he remembered everything, and could correlate information from many areas into a single idea. Woolley's reaction, when Merrifield came up with the idea for solid phase protein synthesis, was crucial:

I worked with him for several years on his projects until I got the idea of solid phase synthesis. I told him the idea. It was kind of funny; we were riding up the elevator and he got off without a word. Next day he came in to see me and said, 'that may be a good idea, why don't you work on it?' That made all the difference.[36]

Arthur Kornberg (M59)[37, 38] shared the Nobel Prize with one of his former mentors, Severo Ochoa (M59), for discoveries related to nucleic acids. Kornberg's approach and interest were more influenced by Ochoa than might be indicated by the fact that their discoveries were made independently. Here is how Kornberg remembers his mentors:

Ochoa was a Spanish refugee who then had little reputation. But he was a very serious scientist. He had no overt political leanings, but his family and friends were all Republicans. He had gone to Germany to do biochemistry and when Germany was no longer hospitable he went to England. With the onset of World War II, he found a place, in 1942, in St. Louis, in the laboratory of Carl Cori (M47). . . . Carl Cori, one of my heroes, made a place for Ochoa, Herman Kalckar, Luis Leloir (C70), and many others. Ochoa then moved on to be a guest at New York University in the Biochemistry Department. He was generous in taking me on when I didn't know ATP from ADP and knew nothing about enzymes. He gave me my start and opened up a new world.

Ochoa had a remarkable imperturbability. Having been knocked around by political upheavals in Spain, Germany, and England, by now in his forties without a laboratory of his own he still had an air of detachment and complete devotion to science. He enjoyed music, good fellowship, and fine dining. Being with him and working with him day in and day out I felt that I might some day be able to do what he was doing. I was in much greater awe of Carl Cori.

Cori was a man of great intellect with a prodigious command of physiology and medicine, an aptitude for chemical kinetics and chemistry. I didn't know that I'd ever match Cori in his intellectual breadth and capacity, and, maybe, I haven't. . . . Gerty liked me and helped me. She was a most remarkable scientist and woman but died in 1957, at the age of 61 of anemia. In the later years Cori mellowed a lot and we grew much closer.[37]

Paul Berg (C80) was a postdoctoral associate with Arthur Kornberg at a time when Kornberg was already a big name. It helped Berg's career enormously that Kornberg let him publish his first important results all by himself. Berg says, 'Had he put his name on the paper, as was traditional, he would've received all the credit and I would have been seen as a promising young student in his lab. As it was, right from the very beginning it was my discovery. I always remembered that it was an incredibly important happening because it provided national recognition.'[39]

Max Perutz (C62)[40] came to Cambridge to study with J. D. Bernal, and was with him for two years. Then Bernal became professor of physics at Birkbeck College in London, but as Perutz was enrolled as a research student in Cambridge, he stayed there and Lawrence Bragg became his supervisor. Perutz ascribes his scientific success to Bragg's support, and to few other careers may the expression 'he stood on the shoulder of giants' be more appropriate.

Alfred Gilman (M94) did his PhD training with Theodore Rall who had been working with Earl Sutherland (M71). Gilman can trace back everything in his scientific career to Sutherland's work, and before that to Gerty and Carl Cori (M47). Gilman then did postdoctoral work with Marshall Nirenberg (M68), whom he remembers as one who 'loves science perhaps like no one else I've ever seen. He's wholly immersed in science to the exclusion of everything else'.[41]

Robert Curl (C96) had two mentors who were not Nobel prize-winners but whose life achievements equaled those of many laureates. He went to Berkeley to be Kenneth Pitzer's graduate student in 1955. He did so because one of his former professors had spoken so enthusiastically about Pitzer. Curl learned more than just science from Pitzer:

> Pitzer's idea of directing graduate students was to leave them alone, but if the graduate student sought him out and asked him for advice, he always made time for them, he was always willing to discuss the problem, offer advice, suggest a solution, and so on. This was ideal for me. He was real busy then because he was the Dean of the College of Chemistry . . . I was always amazed that I never had any trouble getting in to see him. There was another thing he would do for me. I'm one of those people who don't know how to get away when a conversation is over. It was wonderful with Pitzer because we would discuss what I wanted to talk about, we would reach a conclusion, and somehow or another I would find myself outside the door of his office with a warm feeling. I appreciated this efficiency. I was really happy that he was mentoring me in exactly the way I wanted to be mentored.[42]

Then Curl continued at Harvard as a postdoctoral fellow with E. Bright Wilson (Kenneth Wilson's (P82) father):

> In personal styles, these two great men were remarkably different. Pitzer always seemed very relaxed. He approached science in terms of seeing the big picture. He wanted to get at the nub of the matter. His favorite adverb was 'essentially'. In contrast, Wilson believed in taking life quite seriously. He seemed to me to be the quintessential New Englander, a man of rock-ribbed integrity and deep moral sense, even though I knew he was born in Tennessee. He rarely put his name on the papers that came out of his laboratory: his name went on a paper only if he felt he had made a major intellectual contribution to the work. He abhorred superstition in any form; no student or postdoc would dare mention the gremlins we believed lived in his spectrometer in front of him. From what I've said, you might draw the impression that Wilson was a stern, severe person. In fact, he was one of the kindest people I've ever known. He rarely volunteered advice and counsel, but, if you asked for it, he gently provided great wisdom and insight.[42]

One of the main attractions for Sheldon Glashow at Harvard in 1954 was Julian Schwinger 'who was as godlike to me as Michael Jackson is to my children'.[43]

Mario Molina (C95) had two guides. For his doctoral work, George Pimentel was his supervisor at Berkeley. Pimentel was an outstanding scientist who could have won the Nobel Prize for his chemical laser and pioneering matrix isolation studies in spectroscopy. To Molina, another deciding factor was that Pimentel got close to his students socially.[44] From Berkeley, Molina went to Irvine, another campus of the University of California, where he joined Sherwood Rowland (C95). Rowland let Molina choose his project from among several possibilities and supervisor and postdoc mutually increased their interest in atmospheric chemistry. Molina had again landed a mentor who operated in close contact with him.

Hermann Staudinger (C53) received his Nobel Prize rather belatedly for his research on macromolecular chemistry. He was mentor to Leopold Ružička (C39) and Tadeus Reichstein (M50). They both did their doctoral work with him at the University of Freiburg, Germany.[45] Georg Wittig (C79) moved to the University of Freiburg in 1937 and Staudinger may have become a mentor figure for him. Wittig worked on phosphorus chemistry, a subject in which Staudinger had done important research. Decades later Wittig was to produce a publication entitled 'Variations on a theme by Staudinger'. Staudinger was also an anti-Nazi whose moral standing may have provided an important example for Wittig.

It is always fascinating to see people evolve from being excellent pupils to great mentors. Eugene Wigner (P63) was an outstanding example. Once a pupil of Michael Polanyi, Wigner served as guide for several major figures in condensed matter physics in the United States. John Bardeen (P56, P72) was one of them. Bardeen is the only person to have ever received two Nobel Prizes in physics. The first, in 1956, he shared with William Shockley and Walter Brattain, 'for their researches on semiconductors and their discovery of the transistor effect'. The second, in 1972, he shared with Leon Cooper and John Schrieffer, 'for their jointly developed theory of superconductivity, usually called the BCS-theory'.

Bardeen had a large number of Nobel laureates or future Nobel laureates as mentors, and was in contact with more in his early career. When he was a student in electrical engineering at the University of Wisconsin, Bardeen took extra courses in mathematics and physics, and was introduced to quantum mechanics by John Van Vleck (P77). He attended the lectures on kinetic theory and statistical mechanics by P. Debye (C36) who was a visiting professor. He also attended the lectures of other visitors like Paul Dirac (P33), Werner Heisenberg (P32), and Arnold Sommerfeld. Then Bardeen became a graduate student at Princeton University, where there was a close interaction between the professors and fellows of the Institute for Advanced Study, and there were numerous visitors. John Neumann, Herman Weyl, Einstein (P21), and Wigner were among those Bardeen got to know there. Wigner suggested his thesis problem and he attracted many other students and postdoctoral fellows in solid state physics, among them F. Seitz and C. Herring. Herring was later Philip Anderson's (P77) mentor. Following his PhD, Bardeen moved to Harvard for a while where he had a great deal of contact with Percy Bridgman (P46), and Slater at the Massachusetts Institute of Technology; and he met Shockley, one of his future co-recipients of the Nobel Prize. In 1938 he moved to the University of Minnesota and in 1941 he entered war service. Bardeen considered himself 'fortunate to be closely associated with all three of the pioneers in solid state theory in the U.S.A., Van Vleck, Wigner and Slater, as well as with many of the leaders of my own generation, such as Seitz, Herring and Shockley'.[46] From Bardeen's description, Wigner emerges as a popular mentor.

Wigner dedicated almost his entire speech at the Stockholm City Hall to his own mentors.[47] Man's knowledge has become man's knowledge rather than individual knowledge because people can communicate their knowledge and teach each other, he said. He wanted to call attention to how much of our attitude we owe to our teachers. In addition to his high-school mathematics teacher in Budapest (p. 121), he singled out Michael Polanyi, his 'doctor-father' in Berlin, whose teachings remained the guiding principles for Wigner throughout his career. Finally, Wigner stressed how much he had learned from his peers and his pupils. I knew Wigner and I was slightly irritated, as were many others, by his over-polite style. However, his brief Stockholm speech struck me as a most genuine expression of his feelings.

Polanyi acted as a mentor to Melvin Calvin (C61),[48] too. Calvin studied the quantum mechanical theories of reaction mechanisms during his graduate work, and found out that Polanyi, in Manchester, was just developing the best approach to it. For postdoctoral work Calvin went to Polanyi, who introduced him to the porphyrin structures and the importance of the movement of electrons and protons in them. Chlorophyll in green plants is a close relative of porphyrin and also involves photochemical oxidation/reduction. That is how Calvin got started in that area of research, which led eventually to the Nobel Prize 'for his research on the carbon dioxide assimilation in plants'.

In an indirect way, Michael Polanyi influenced yet another future laureate, his son John (C86):

> Most of what he taught me about physical chemistry I learned at one remove from him. I was a student for six years in the department that he had shaped in Manchester. The professor was one of his favorite students, Meredith Evans, and my PhD supervisor was another student of his, Ernest Warhurst. What I learned from his students gave me a sense of scientific values— where the field was going, what were the important questions to tackle, and, to a degree, how to tackle them. Without those things I would have been lost. But it happens that I didn't get them directly from him, but from people who owed a lot to him.[49]

A casual remark

We conclude with an example of how a great teacher's almost casual remark can become a life lesson and motivation for a young disciple's most distinguished career. Frank Westheimer[50] of Harvard University is one of those non-Nobel laureates who could have won. He happened to have an adviser at Harvard who never told him anything and was barely around during his graduate studies. Westheimer had originally come to Harvard in 1933, hoping to become a student of the organic chemistry professor, James B. Conant. However, Conant had just become president of Harvard and Westheimer had to sign up for someone else. (Conant's career later took him to impressive heights in public life. During the Second World War he served as chairman of the United States National Defense Research Council, and after the war at one time he was the American commissioner for Germany.)

As Westheimer was graduating from Harvard in the mid-1930s, President Conant remembered him:

About that time, Conant called me into his office. He said that he knew I was getting my doctorate, and was interested in my career. What was I going to do? I told him I'd won this fellowship, and explained with great pride the problem I'd submitted and was going to work on. Conant had the habit of putting the tips of his fingers together and rocking back and forth while he thought, and he put the tips of his fingers together and rocked back and forth, and then he said, 'Well, if you are successful with that project, it will be a footnote to a footnote in the history of chemistry.' As I walked out of his office, I realized what he had told me.

Really, it was two things. One was, of course, that my project wasn't very important. The other was—and it may have been pretty stupid that I had never thought this until that moment—that I was supposed to do important things. Chemistry was a lot of fun; it was great entertainment, and I was going to be paid—or at least so I hoped—for entertaining myself with it. Yet Conant had essentially told me that I was expected to do things that were scientifically important. The interview with Conant provided a vital kick in the pants for me. It changed the way I thought about my future.

At Columbia, I did the project that I had proposed, and it worked out beautifully. But it was exactly what Conant had said it was, a footnote to a footnote in the history of chemistry.

Then I set my sights higher—much too high, as a matter of fact. As a physical–organic chemist I was concerned with general acid–general base catalysis, and had decided that enzyme catalysis was probably caused by simultaneous general acid–general base catalysis. I was going to demonstrate this in my next piece of research. . . . The project was obviously enormously ambitious, and although I was fundamentally correct about enzymes, demonstrating it was much too big a project for me at the time. That attempt came to nothing, but at least Conant wouldn't have been able to object that the attempt was directed at a footnote to a footnote.

Eventually I settled down to things that were more important than footnotes to footnotes, but not as grandiose as the youthful project that I just mentioned. I never discussed my research with Conant again, but I did restrict myself to things that he might have approved of.

Many years later, after I was a professor at Harvard, and after Conant had retired from his many careers, I was working in my office one Saturday when someone knocked on my door. I opened it, and there he stood. He looked at me and said, 'Do you remember me?' Needless to say, I did.[50]

9

Changing and combining fields

The changes in research areas by great scientists often come from the attraction of a new idea or challenge, or the emergence of a new tool. Such a change may be gradual, with the old topic lingering around for a while as the new is getting introduced, or more dramatic, as when an outside impulse or an internal recognition makes the scientist abruptly abandon a successful area for a new field. We can compare it to a master detective getting a fresh case: he sweeps clean his desk in a matter of hours and switches to his new task. Future Nobel laureates and other great scientists change the subjects and techniques of their research in a great variety of ways. What is common is the willingness and ability to change. Biochemist Erwin Chargaff (see p. 58) is a beautiful example of a scientist who moved from a field in which he was very successful. Under the impact of the publication by Avery, MacLeod, and McCarty in 1944,[1] in which DNA was shown to be the substance of heredity, Chargaff switched to the study of nucleic acids. Seldom do scientists so eloquently describe such a transfer:

> Consequently, I decided to relinquish all that we had been working on or bring it to quick conclusion, although the problems were not without interest and dealt with many facets of cellular chemistry. I have asked myself frequently whether I was not wrong in turning around the rudder so abruptly and whether it would not have been better not to succumb to the fascination of the moment; but these biographical bagatelles cannot be of interest to anybody. To the scientist nature is as a mirror that breaks every 30 years; and who cares about the broken glass of past times?[2]

Innocent gaze

Often the change in topics or research area means that the scientist leaves an area in which he is experienced to enter another in which he is a relative beginner. The advantage of being an outsider in a field has been noted repeatedly. William Astbury characterized his approach to accumulating information from his X-ray diffraction experiments in this way: 'the panorama unrolled itself to my *innocent* gaze'.[3] He also believed that 'the greatest asset of scientific research is its *naïveté*'.[3] According to Gerald Edelman (M72), '*Naïveté* is often very important when you are young as long as you are knowledgeable.' Then he added: 'Another thing that is important is to find a contradiction in the literature.'[4] Thus being innocent and naïve and being merely an outsider do not suffice, one also needs to be an insider and knowledgeable in another field.

Francis Crick was about 30 when, after the years lost due to military service, he was looking for a direction in his life. Basic research attracted him, and though he was inexperienced, he gradually realized that his

> lack of qualification could be an advantage. By the time most
> scientists have reached age thirty they are trapped by their own
> expertise. They have invested so much effort in one particular field
> that it is often very difficult, at that time in their careers, to make
> a radical change. I, on the other hand, knew nothing, except for a
> basic training in somewhat old-fashioned physics and mathematics
> and an ability to turn my hand to new things.[5]

In this sense Crick had a free choice. Crick and Watson's joint work on the structure of DNA required an intellectual investment in the study of genetics, biochemistry, chemistry, and physical chemistry, including X-ray diffraction. This was a tremendous scope to cover, and although they have been criticized for not mastering them all perfectly, it was not for want of trying.[6]

Combining the experience of different fields often proves fertile, and may be exceptionally beneficial for making a discovery. In 1933 Jerome Karle (C85)[7] entered City College of New York, where there were broad course requirements for all students, from mathematics to physical sciences, to social sciences and literature. They even had a compulsory public speaking course. Karle studied more mathematics and physical sciences than required, and following graduation he earned a Master's degree in biology. His first contribution to science was an entirely practical project: he developed a procedure for determining the fluorine content in drinking water for the New York State Health Department in Albany.

During his graduate studies at the University of Michigan, Karle learned the gas-phase electron diffraction technique of molecular structure determination. He joined the Naval Research Laboratory in Washington in 1946, and at first worked on developing the quantitative aspects of gas-phase electron diffraction. In this work he came to solutions that had implications for crystal structure analysis by X-ray diffraction. Then he and Herbert Hauptman (C85),[8] a mathematician turned crystallographer, developed the so-called direct method of crystal structure analysis, which greatly broadened the possibilities of X-ray crystallography and earned them the Nobel Prize. In an X-ray diffraction experiment half of the information is lost, or at least that used to be the accepted dogma in X-ray crystallography. It took the outsiders Hauptman and Karle, coming from mathematics and gas-phase electron diffraction, respectively, not to be bogged down by this dogma. Their team had a double advantage in tackling the phase problem: in addition to being relative outsiders in the field, they combined their different areas of expertise in a most fortunate way.

Aaron Klug (C82)[9] has more than once used the experience gained in one field fruitfully in another. Originally with a PhD degree in physics, Klug started out in electron microscopy with the simple view that it was well understood. He soon realized, however, that this was not the case. Klug was lucky that he was a non-expert and not full of presuppositions. There were various people working in the field trying to get 'the perfect picture' and he realized that there was no such thing. He introduced the approach of taking a series of micrographs in the electron microscope, at various degrees of defocusing, and then correcting them. With this method it became possible to create an image of transparent objects, that is, non-stained biological specimens. The technology further developed in the course of practical studies. They started out with a real problem of a helical virus and soon realized that they could make a three-dimensional reconstruction by using the theory of helical diffraction. Klug had developed the approach for Rosalind Franklin's X-ray studies of the tobacco mosaic virus. Later he saw that it was a special case of a more general principle in Fourier theory.

Klug's method also became the basic principle of the X-ray CAT (computer assisted tomography) scanner for which Godfrey Hounsfield (M79) and Allan Cormack (M79) received the Nobel Prize. Klug had realized earlier that his method could be applied to medical radiography, and talked to radiographers who brushed off his 'fancy stuff'. They said, 'We understand exactly what we see in a medical X-radiograph.' They did not realize that they were using a much bigger dose of X-rays than necessary. Later it was shown that CAT-scanning image reconstruction could provide more information for a given dose than X-ray tomography, and CAT scanning has now become the standard procedure. To Klug 'This story illustrates the important point in science that you sometimes find the solution to a problem from another field.' Under more fortunate circumstances and with more receptive partners, Klug could have shared the Nobel Prize for CAT.

Cross-fertilization

The willingness to change is applicable to Lars Ernster's definition of both digging and drilling scientists (see p. 61). The correspondence to the digging type is more straightforward, but a drilling type also needs to change techniques and methodologies as new ones come along, and may involve the achievements of other fields in his research. Aaron Klug (C82)[9] belongs to the digging category, but he immersed himself so deeply in each new field that, taking each one separately, he also qualifies for the drilling classification. He worked in five areas: the structure and assembly of the tobacco mosaic virus; spherical viruses and electron microscopy; the structure of transfer RNA and later an RNA enzyme (ribozyme); chromatin and the nucleosome; and zinc finger proteins.

Klug's career provides prime examples of the cross-fertilization of experiences. The first two topics were mentioned above. His PhD work in the Cavendish Laboratory was on a problem of kinetics of phase transition in solids. There he used a concept of the nucleation of the new phase. Later, while thinking about the tobacco mosaic virus assembly, he was helped by this prior experience in recognizing that he had to separate nucleation and growth.

When studies of protein synthesis were at the forefront of molecular biology and the genetic code was an important issue, Francis Crick (M62) turned Klug's attention to the structure of transfer ribonucleic acid (tRNA). In one loop of tRNA there is a codon that recognizes the code of the messenger RNA, which is the three bases specifying the amino acid. At the other end of tRNA there is the site of attachment of the amino acid.

I *The Nobel Prize and Sweden*

The Nobel Medal for
physiology or medicine.

Announcing the Nobel Prize in Chemistry for the discovery of fullerenes at the Royal Swedish Academy of Sciences on 10 October 1996. From left to right, Carl-Olof Jacobson, secretary of the Academy, Salo Gronowitz, chairman of the chemistry committee, and Lennart Eberson, a member of the chemistry committee.

Nobel Prize ceremony in 1978. In front from left to right, Prince Bertil, King Carl XVI Gustav, Queen Sylvia, Princess Lilian. Second row, representatives of the Nobel committees; from left to right, Lars Werin, economics, Lars Gyllensten, literature, Peter Reichard, physiology or medicine, Lars Ernster, chemistry, Lamek Hulthén, physics, and Sune Bergström (M82), chairman of the board of directors of the Nobel Foundation.

2 The Nobel Prize and national politics

British Nobel laureates at the British Embassy in Stockholm in 1991 during the 90th anniversary of the first awarding of the Nobel Prize. Sitting from left to right, Dorothy Hodgkin (C64), Godfrey Hounsfield (M79), Max Perutz (C62), Aaron Klug (C82), Andrew Huxley (M63), Frederick Sanger (C58, C80), George Porter (C67), César Milstein (M84). Standing, left to right, Antony Hewish (P74), Bernard Katz (M70), Geoffrey Wilkinson (C73), William Golding (Literature 1983), James Black (M88), and John Vane (M82).

Richard Kuhn (C38).

Adolf Butenandt (C39).

3   Who wins Nobel Prizes?

Kary Mullis (C93), 2000.

Linus Pauling
(C54, Peace 1962).

William Astbury.

Murray Gell-Mann (P69) (second row, first from the right) and Yuval Ne'eman (first row, first from the left) at the Eighth Nobel Symposium in Göteborg, 1968. Abdus Salam (P79) is second row, third from the left.

James Watson (M62) measuring a model of DNA in Cambridge, early 1950s.

Erwin Chargaff in New York City, late 1940s.

Stanley Prusiner (M97).

Charles Weissmann in
London, 2000.

The authors of the paper that reported
the discovery of buckminsterfullerene, C^{60}.
Front, left to right: J. R. Heath, R. E. Smalley (C96),
H. W. Kroto (C96), and S. C. O'Brien; back: R. F. Curl
(C96). Taken in September 1985 at the end of
the week during which the discovery was made.

Maria Goeppert-Mayer (P63) in 1949.

4 Discoveries

Albert Einstein (P21) in 1921.

Max Planck (P18).

Albert Szent-Györgyi (M37) and Dorothy
Hodgkin (C64) in New York City in the late
1940s.

Tsung-Dao Lee (P57) and Chen Ning Yang (P57)
visiting a physics laboratory in Stockholm, 1957.

Frederick Sanger (C58, C80) at the time of his
first Nobel Prize.

Marshall Nirenberg (M68) in
Bethesda, Maryland, 1999.

5   *Overcoming adversity*

Henry Taube (C83) on the day his Nobel Prize was announced.

Rosalyn Yalow (M77) in 1948.

Petr Kapitsa (P78) in Cambridge, late 1920s.

Lev Landau (P62) in Moscow, 1937.

6 *What turned you to science?*

Eugene Wigner (P63) and the author in Austin, Texas, 1969.

James Black (M88) in London, 1998.

Carleton Gajdusek (C76) with children on Tongariki Island in the New Hebrides. The island disappeared in violent earthquakes in the 1960s.

7 Venue

Francis Crick (M62) and James Watson (M62) in Cambridge, about 1950.

Joseph Thomson (P06) and Ernest Rutherford (Co8) outside the Royal Society Mond Laboratory in Cambridge, 1933.

Niels Bohr (P22).

Rita and John Cornforth (C75).

8   *Mentor*

Viktor Hamburger and Rita Levi-Montalcini (M86) in St. Louis, Missouri.

César Milstein (M84) and Frederick Sanger (C58, C80) in Cambridge.

Severo Ochoa (M59).

Sherwood Rowland (C95) and Mario Molina (C95) in Irvine, California.

Frank Westheimer.

James Conant.

9 *Changing and combining fields*

George Porter (C67) as a lieutenant in the
Royal Navy, 1943.

Walter Gilbert (C80) in Indian Falls,
California, 1998.

Vladimir Engelhardt in Moscow.

Boris Belousov.

Anatol Zhabotinsky in Waltham,
Massachusetts, 1996.

Oliver Lowry, the most cited scientist in history.

11   *Is there life after the Nobel Prize?*

César Milstein (M84) with his father in Stockholm, 1984.

The author and James Watson (M62) in Budapest, 2000.

Left to right: Edwin McMillan (C51), Glenn Seaborg (C51), John Kennedy, Edward Teller, Robert McNamara, and Harold Brown in Berkeley, California, 1963.

Werner Heisenberg (P32), 1949.

12   *Who did not win*

J. D. Bernal in London.

Oswald Avery.

Maclyn McCarty being congratulated by James Watson (M62) and Francis Crick (M62) upon receiving the First Waterford Biomedical Award in La Jolla, California, 1977.

Jonas Salk.

Albert Sabin.

Neil Bartlett in Vancouver, around 1962.

Otto Hahn (C44) and Lise Meitner at the Hahn Prize presentation to Lise Meitner in 1955.

Antony Hewish (P74) in Cambridge, 2000.

Jocelyn Bell in Princeton, 2000.

Donald Huffman and Wolfgang Krätschmer in Huffman's laboratory at the University of Arizona, Tucson, 1999.

Salvador Moncada in London, 2000.

Although Klug was very successful in his virus studies, he knew he did not want to stay in the same field for too long and switched to researching the tRNA structure.

Experimental research needs a lot of support and it takes a far-sighted view from the funding agencies to back a scientist when he wants to leave a successful area of investigation for a new and risky one. Having solved what he had set out to solve about the tRNA structure, Klug decided to abandon this highly successful field and move on again, to study chromatin, which is a complex structure of DNA and protein molecules making up the chromosomes. His intention to switch surprised his referees on a visiting committee from the funding agency, who gave Klug a black mark for it. Klug also abandoned, at some point, electron microscopy, which he had established as a major addition to structural biology.

Securing the necessary funding for research is a major consideration, especially in experimental science. Few funding agencies provide unrestricted grants, and Peter Debye's (C36) remark still holds true: 'it is not easy to get money for a thing which is wild—where you cannot say "this is going to have results." '[10]

It is a temptation, and a danger, to stay with successful experiments and methodologies for too long. The situation may be compared to the movie *Ground Hog Day* in which the main character relives the same day over and over, with all the benefits and frustrations of the experience. Sherwood Rowland (C95)[11] learned from his PhD supervisor, Willard Libby (C60), that the excitement and fun in science comes from doing new things. Rowland likes to explain this with two seemingly similar expressions from English. One is 'in the groove', which means that things are going very well. The other is 'in a rut', which has physically the same meaning but actually signifies that you have trapped yourself into doing the same thing over and over again. The two are not always easy to delineate. Doing something really new means being in the groove, but one may pound on it for too long and slide into a rut rather easily. First experiments induce a lot of excitement because they are excursions into the unknown. By the time it is going smoothly and predictably, in principle it is no longer an experiment worth doing because the experimenter knows in advance what the results will be. Rowland was always conscious of staying in the groove long enough but getting out before finding himself in a rut. He took a new direction every six to ten years.

Donald Cram (C87)[12] was another scientist who felt that retaining his fascination with science required a periodic change in research areas; for him it was about every ten years. Cram had the long-range ambition of modeling aspects of biological systems with designed simpler synthetic organic systems. In the early seventies Cram borrowed a large set of molecular models, and during the first three days of playing with them he got so carried away that he never got dressed. He constructed a large variety of potential host–guest systems with the models. His new area was not only fascinating, it was also pregnant with the possibility of failure. Cram did not withdraw into the seclusion of his laboratory to see whether his new ideas would work or not; rather, he talked about them to all who would listen as if challenging himself to going through with the changes in his research. At first he even found it difficult to convince his own graduate students to join him in his new endeavor. After he had pronounced his new ambition to his colleagues internationally, there was no way back; he had to charge ahead, and he succeeded spectacularly.

Shifting interests

Jean-Marie Lehn (C87)[12] started as an organic chemist. Over the years he has become interested in a number of other areas, especially in physical chemistry, including molecular physics of liquids and motion in liquids. His studies of selective binding of metal ions led first to molecular recognition, and he then expanded this into the general concept of supramolecular chemistry. According to Lehn, this change of interest was linked more to a frame of mind than to an outside influence. At one stage he wanted to study philosophy, and he always liked to put what he was doing into a more general framework. Concepts inspired him, made him think more broadly, and made him look at the objects he was studying in a more general fashion.

Kenichi Fukui (C81)[13] started as an experimental chemist, and wrote more than 100 papers on the subject. It was by chance that he became interested in the chemistry of hydrocarbons and, as a consequence, in theoretical work. He had become engaged in the chemical conversion of olefinic hydrocarbons after his graduation, and had taught himself quantum mechanics. After that, his theoretical work gradually surpassed his experimental work.

Manfred Eigen (C67)[14] received his Nobel Prize at age 40, for research in the extremely fast reactions that he had carried out at an even younger age. Then he changed fields rather drastically, to the slow evolutionary processes. He stresses, though, that evolution did not take billions of years, as many people believe. It took humans such a long time to develop, but not the first cell. Employing dating experiments that use RNA data, Eigen and his co-workers have found that the genetic code already existed 4 billion years ago. Since the earth is 4.5 billion years old and since it had to cool down first, there was not much time for all that chemistry to come about. Then it took about another 3 billion years to get from a single cell organism to a human.

John Bardeen (P56, P72) and Frederick Sanger (C58, C80)[15] each won two Nobel Prizes in their respective fields, so, obviously, they changed topics between the first and the second. Bardeen's first Nobel Prize was for the transistor, together with his two colleagues at Bell Laboratories. He then gradually returned to his earlier interest, the theory of superconductivity. He felt that this would be more appropriate to pursue in an academic environment and moved to the University of Illinois. Thus he changed not only research topics but venue as well.[16] After the first prize, Sanger initially continued to study proteins. His prize-winning work had been in sequencing proteins so the question of sequencing nucleic acids came up. In 1958 nothing was known about sequencing nucleic acids and it did not seem to Sanger an easy problem, for two reasons. One was that the nucleic acid molecules were so big and the other was that they had only four components. To work out a sequence with only four components would be very much more difficult than the sequence of a protein with 20 components, the amino acids. Sanger, however, realized that it was a very important problem. As DNA seemed quite impossible, he started thinking about RNA. The transfer RNAs were small, of about 70 or 80 nucleotides, and around 1965 he started working on RNA.

Paul Berg (C80) had become a successful scientist very young, starting with Arthur Kornberg (M59) and quickly establishing himself as an independent researcher. In 1967 he left his university for the Salk Institute in La Jolla, and worked with Renato Dulbecco (M75). Dulbecco had only recently discovered a new group of viruses called tumor viruses. Following infection of cells, these viruses can change the cell from being normal to cancerous. When these viruses were injected into mice, they developed tumors. Rather than killing the cell, the virus converts it into a cancer cell. It was virus genes that made the cell grow like a cancer cell. Berg wanted to find out the mechanism of all this. At this point he decided to change the whole direction of his research to work on that problem. He established a new kind of laboratory, and dropped what he had been doing, which had been extremely successful. In Berg's words, 'I was certainly a leader in the field in which I was working at that time. But it seemed to me that that was going to go on and on and on the same way, and the prudent thing was to ask some new questions.'[17] His was another example of a scientist's conscious decision based on the distinction between being 'in the groove' and 'in a rut'.

After Albert Szent-Györgyi (M37) had won his Nobel Prize, he was faced with the interesting question of what to do next. He could have continued his prize-winning research but he felt he had done what he wanted in that area. While studying the literature a protein called myosin, an important component of muscle cells, caught his eye. It seemed to him under-investigated, and it was also at a time when important discoveries on ATP were taking place. Then in 1939 a Soviet couple, V. A. Engelhardt and his wife, M. N. Lyubimova, published a paper in *Nature* stating that myosin is an enzyme and that it catalyzes the energy production from ATP. This immediately sparked Szent-Györgyi's imagination and he moved into muscle research with the enthusiasm of a beginning scientist.[18]

Saying that Luis Alvarez (P68)[19] changed areas of research is an understatement. He had an extraordinary spectrum of activities in physics, including applications. He established the first ground control approach for landing airplanes, and all American planes were equipped with Alvarez's ground control lander during the Berlin airlift in 1948. He also invented the hydrogen bubble chamber, which won him the Nobel Prize. Later he initiated the theory that dinosaurs had been wiped out by a large comet or meteor, which his son Walter then worked out in greater detail. According to the theory, there was so much dust in the atmosphere following the impact that the Earth was in darkness, with a resulting lack of food for the dinosaurs.[20]

Harold Kroto (C96) decided to study spectroscopy at Sheffield University as an undergraduate. Then he did a PhD in spectroscopy and subsequently became interested in microwave spectroscopy for the determination of the structure of small molecules. Around the early seventies the tremendous breakthrough by Charles Townes (P64) and colleagues occurred in the detection of interstellar molecules. Kroto, like many other spectroscopists, felt that he should dive into this field:

> If you are a scientist and if you have a technique that is absolutely
> perfect, and one that is as complicated as microwave spectroscopy
> is, then you could really make a synergistic complementary
> contribution if you worked with a radio-astronomer, who knew
> where the sky was. They could not overnight learn the theory that
> was required to understand spectroscopy.[21]

Kroto combined his interest in interstellar molecules with Richard Smalley's (C96)[22] cluster experimental technology in suggesting an experiment to study the evaporation products of graphite. The idea that such a combination should work came from a third person, Robert Curl (C96),[23] who brought the other two together. Experience in different fields came in handy even in naming the highly stable and symmetrical C_{60} molecule. Both Kroto and Smalley, in their own ways, associated their model with Buckminster Fuller's Montreal geodesic dome, and invoking the architect in the name buckminsterfullerene also meant invoking the whole science of design, and very appropriately because the shape of the truncated icosahedron is beautiful in itself.

Crossing the lines

Gerald Edelman (M72)[4] took a medical degree originally, but then trained in the physical chemistry of macromolecules and received a PhD at the Rockefeller Institute. He determined that the antibody molecule had chains, which made it easier to analyze its sequences. The fact that it consisted of chains also had important implications for its medicinal chemistry. Edelman eventually determined the structure of the antibody molecule by going from medicine to chemistry, back to genetics, and then returning to chemistry and stereochemistry. It was an outstanding example of a confluence of special talents.

George Porter (C67) served in the Royal Navy in the Second World War. This experience proved beneficial in his research when he incorporated the huge capacitors acquired as navy surplus in his kinetics experiments. He said, 'There is nothing more rewarding than linking two quite different subjects.'[24] In fact, the idea for his research came by combining experience of radar and photochemistry. There was interest at the time in free radicals, which lived for a thousandth of a second or less. Many people even doubted their existence, and Porter wanted to see them. Instead of using a continuous source of light, an army search light, for example, he used the very powerful flash lamps that had been developed during the war. These had been installed in bombers that were flying over Germany taking photographs, the bomb bays being filled with condensers to store the energy. Porter acquired most of the parts for his experiment from navy surplus. The second idea was to use two flashes, one as a pulse and the other as a probe. This was the optical equivalent of radar, which uses pulses of electromagnetic radiation, radio waves. These pulses were a few millionths of a second long, compared with the microseconds of pulses in the latest navy radar sets. The idea was to send out one pulse and then probe it, using the velocity of light to change distance into time. So Porter acquired not only equipment but also ideas from his navy service.

Gertrude Elion (M88) cultivated many fields in her drug research at Burroughs Wellcome. Her colleagues called her department a 'mini-institute' since it contained sections on chemistry, enzymology, pharmacology, immunology, virology, and a tissue culture laboratory.[25]

Ilya Prigogine (C77)[26] has a background in chemistry and physics. The main conclusion of his studies on non-equilibrium thermodynamics is that the direction of time is of fundamental importance in nature. Classical mechanics, even quantum mechanics and relativity, are time-symmetrical theories, the past and the future playing the same role in each of them. On the other hand, thermodynamics introduces entropy, the measure of disorder, and entropy is associated with the arrow of time. This leads to questions about the role of entropy and of the distance to equilibrium in nature and about the relationship of the time-symmetry breaking of entropy to the laws of physics. A near-equilibrium world is a stable world in which fluctuations regress and the system returns to equilibrium. The situation changes dramatically far from equilibrium. Here fluctuations may be amplified and, as a result, new space–time structures arise at 'bifurcation' points. Prigogine introduced such concepts as self-organization and dissipative structures. Irreversible processes associated with the flow of time have an important constructive role. Prigogine's dual background was beneficial in his work of incorporating the direction of time into the fundamental laws of physics. Originally Prigogine had been more interested in philosophy than in science, which helps him understand the gap between philosophical and scientific thinking.

One of the sources of Linus Pauling's (C54, Peace 1962) greatness was the ease with which he crossed disciplinary lines. He was the first chemist to understand quantum mechanics and apply it to chemistry, thereby creating the initial understanding of the nature of chemical bonding. He was always alert, ready to absorb new knowledge and apply it to new areas. A lecture he presented at the Rockefeller Institute in 1936 gave Karl Landsteiner (M30), the discoverer of human blood groups, an opportunity to ask him about the molecular basis of antigen–antibody reactions. This interaction prompted Pauling to look even more generally into the question of biological specificity, and to arrive at the concept of complementariness in associations of biological macromolecules.[27]

Pauling's broad interests were legendary, and some were much criticized. In particular, many were disappointed by the lack of evidence to back his claims for the beneficial effects of vitamin C and the extraordinarily large doses that he advocated. This was later in his life and may have come from, by then, being insufficiently versed in areas that he was trying to incorporate into his arsenal of tools.

Kary Mullis (C93) received his Nobel Prize 'for his invention of the polymerase chain reaction (PCR) method'. He had thought about becoming a physicist, but he 'knew that as soon as the Russians and the Americans decided that they were not at war with each other, or one of us was overcoming the other, they won't need as many physicists as they used to, and this is exactly what happened.'[28] While at Berkeley as a graduate student in biochemistry, he swung back and forth between physics and biochemistry. He 'thought it was a good idea to go into biochemistry because politicians are always getting heart attacks, and they die of cancer, so they'll need us. But they won't care what's falling out of the sky and they sure don't care how old the universe is.' However, his first academic paper, which he published by himself as a graduate student in biochemistry, was about the deep secrets of the universe. It appeared in *Nature* and he called it 'The Cosmological Significance of Time Reversal'.[29]

Sidney Altman (C89)[30] originally studied physics, and was in Colorado in the summer of 1962 to attend a summer institute in theoretical physics. There he met George Gamow, a famous and flamboyant physicist, who told Altman about interesting work in molecular biology in the medical school. When Altman transferred to molecular biology, he had first to spend a year learning biochemistry and organic chemistry.

John Pople (C98) is a mathematician turned chemist. He was trained in Britain, where it is not uncommon for mathematicians to go into theoretical physics and he carried this just one step further, into theoretical chemistry. He said 'I see myself as a chemist although I have no professional qualifications in chemistry.'[31] Walter Kohn (C98) calls himself a 'card-carrying physicist'. His work on the density functional theory is an area of overlapping interest between physics and chemistry.[32] Pople said, at the evening banquet held in the Stockholm City Hall, 'I rejoice in the opportunities I have had in many diverse branches of science over the past fifty years. Starting in mathematics, using fundamental principles of physics, aided by developing power of computer science I have seen the expansion of theory in many branches of chemistry and biology. Increasingly, science is being unified.'

Edwin McMillan (C51) graduated as a physicist, but he took more chemistry than was customary for physics majors and his first publication, in 1927, was a chemistry paper. His last publication appeared in a mathematical journal in 1980 and was co-authored by a mathematician.[33]

Transferability

The ultimate example in crossing disciplinary lines is Walter Gilbert (C80).[34] He has had a remarkable career ranging from theoretical physics to experimental biology, winning a Nobel Prize in chemistry and having been a successful entrepreneur as well. One of his many distinctions is that he was New England entrepreneur of the year in 1991. Gilbert did not merely modify his studies at some point, did not assume the directorship of a company as a side project to augment his professorship, he made full changes, taking the risks and plunging into the new challenges full scale. He did theoretical physics in his graduate studies at Harvard, and at Cambridge where his supervisor was Abdus Salam (P79). Then he returned to Harvard and worked alone, but his roommate was another future Nobel laureate, Sheldon Glashow (P79), and Gilbert also worked as Julian Schwinger's (P65) assistant for a year. Soon, Gilbert was on the faculty of the Physics Department of Harvard, teaching theoretical physics of great diversity, ranging from quantum theory of fields to general relativity, electricity and magnetism, and classical mechanics, in rotation, for about five or six years.

Gilbert met James Watson (M62) in the fall of 1955 and they became friends. Watson already had the double helix discovery behind him and was an assistant professor of biology at Harvard. They talked about what was going on in science and, gradually, Gilbert spent more and more time in Watson's laboratory. He worked with him on the experiments that led to the discovery of messenger RNA in 1960. They faced the problem of showing that there was an unstable RNA intermediate that was copied from DNA and then used to make protein. Until that time it was thought that RNA might make proteins but that the RNA code was stored somewhere inside the ribosomes. Then a line of experiments began to suggest that there might be RNAs that were immediate copies of the DNA.

There was also a series of experiments from the French pair, Jacob (M65) and Monod (M65), suggesting that there could be an intermediate and unstable RNA. Gilbert and Watson showed that there was an unstable form of RNA in bacteria that copied the DNA, went to the ribosomes, was used to make proteins, and was then destroyed. They published the results of these experiments in *Nature* in the beginning of 1961. This was Gilbert's introduction to molecular biology.

As Gilbert continued to work in biology, the Physics Department begin to look very doubtfully at him, and when he came up for promotion it claimed that he was no longer working in physics. However, Harvard created a different appointment, and promoted him to tenure in biophysics. Next he worked on the genetic control mechanism. How does DNA function to make proteins? By that time, the middle 1960s, the genetic code had been broken. Then Gilbert became interested in trying to discover the nature of the control mechanism by which one gene controls another. The leading theory was that of Jacob and Monod, that there were compounds, called repressors, which were the products of one gene that affected and controlled the function of another gene. The mystery at that time was the nature of these repressors. Most of the experiments failed, but later Gilbert and Benno Müller-Hill, his postdoctoral fellow from Germany, devised an experiment that finally succeeded.

Gilbert considers his career to be a message about the usefulness of being educated broadly. The problem is to know enough odd little things about the world. The discovery for which he received his Nobel

Prize, DNA sequencing, depended on knowing, at a certain moment, a certain odd fact about the sugars in DNA. It is difficult to pinpoint how he happened to know that; it may have been an experiment that he had been involved in years before. Gilbert cites Edison, who used to say that being an inventor involves knowing all kinds of apparently irrelevant connections about the world.

According to Gilbert, it is easier to change fields than most people believe. Training that is too specialized may be a handicap. He sees great advantage in the liberal education that the British used to have, of being trained for nothing in particular; that makes one feel free to do anything. Gilbert mentions reading the classics, which does not train you to do anything in particular, but is training in a fundamental way. University of Chicago president Robert M. Hutchins comes to mind; he wanted to base university education in the 1930s on the hundred 'Great Books'. He advocated a college education founded primarily on philosophy, psychology, and literature, with less attention paid to chemistry, physics, and mathematics. Gilbert says:

> There is constant change in the sciences, and if you don't
> realize this, you remain frozen behind. When I changed from a
> mathematical kind of physics to experimental biology, I realized
> that in my work for my physics Ph.D., I learned an essential thing
> that was transferable. That particular ability was how to decide
> myself whether something was right or wrong. That was the crucial
> element, and that ability was transferable to another field. I had to
> learn the particular knowledge of my new field, but one can learn
> the particular knowledge of any field in a short time.[34]

Beyond science

We conclude our discussion of changes with a few examples of scientists who made transitions that led them out of the realm of science, fully or in part, and for whom these other activities led to Nobel Prizes in other fields. These are admittedly extreme cases. Andrei D. Sakharov (Peace 1975) of Russia and Joseph Rotblat (Peace 1995) of Great Britain were both physicists and received the Nobel Prize for Peace. Sakharov had contributed fundamentally to developing the Soviet hydrogen bomb, but in his later years became a relentless critic of Soviet tyranny. Rotblat worked on the Manhattan Project to develop the atomic bomb, but left the project before the first bomb was dropped because he opposed its use after Germany's defeat. He has been very active in the Pugwash Conferences on Science and World Affairs, with which he shared the prize. John Harsanyi (Economics 1994) was originally a pharmacist. He shared the economics prize with John F. Nash and Reinhard Selten, 'for their pioneering analysis of equilibria in the theory of non-cooperative games'.

In 1981, a doctor of chemistry, Elias Canetti, received the Nobel Prize in Literature, 'for writings marked by a broad outlook, a wealth of ideas and artistic power'. Canetti was born in Bulgaria, in a Sephardic family. He wrote all his works in German although he spent most of his life in England. He was a reluctant chemist, although he carried his studies through to his doctorate. Canetti's decision to become a writer was inspired by the lectures of Karl Kraus in Vienna, and this is one more connection, however indirect, to the world of chemistry. We mentioned Erwin Chargaff in the introduction to this chapter. Chargaff has written philosophical books, including his autobiographical volume,[35] and he writes of Karl Kraus:

> He was the deepest influence on my formative years; his ethical
> teachings and his view of mankind, of language, poetry, have never
> left my heart. He made me resentful of platitudes, he taught me
> to take care of words as if they were little children, to weigh the
> consequences of what I said as if I were testifying under oath. For
> my growing years he became a sort of portable Last Judgment.
> This apocalyptic writer . . . was truly my only teacher.[36]

10

Making an impact

Once there is a discovery, the next step is to make it known so that others can use it and, in the final account, mankind can benefit from it. There are different ways for disseminating the results, from a press conference to a scholarly monograph. Holding a press conference as the primary means of communicating new scientific results is suspect because it circumvents the filter system of peer review, and there is seldom a discovery that calls for informing the general public before the scientific community. The most common and time-honored method is to submit an account to a peer-reviewed scientific periodical and to report to a scientific meeting. The whole process of communicating the results may have to start with naming the discovery if it is a truly new contribution.

Power of names

Coining a name is a rewarding, if not a frequent task in scientific research. It is beneficial to designate a new phenomenon or direction in science by a name that will distinguish it from other phenomena and other branches of science. A name is needed for identification; without a name everything is more difficult. In the book, and movie of the same name, *The Day of the Jackal*, the hunt for President De Gaulle's would-be assassin is at first hindered by the lack of a name. To establish the identity of the assassin is more difficult, but without a name it is impossible to begin the hunt.

Aaron Klug (C82), looking back on the beginnings of molecular biology, thinks that it was historically important for the new field to define itself:

> Sometimes you have to define yourself *against* what exists. When I
> was a research student in the fifties, the biochemists never thought
> in terms of DNA transmitting information. They thought of
> proteins as the molecules of life. This is why it was important to

use another name. It is like religion in a way. You had to define the
Protestant religion against the Catholic during the Reformation.
Eventually it will be absorbed in biochemistry.[1]

An important part of the discovery of the third modification of carbon was
giving it the name buckminsterfullerene. As Kroto (C96) explains:

> There is something in Shakespeare's *Othello* (III iii) about one's
> attitude to names. It's a very personal thing. . . . If we had not
> discovered C_{60}, it would have been discovered within a year. In
> fact, it should have been discovered already in the sixties. The
> point is that if I paint a picture, it is the creation of an individual—
> so there is a certain individuality quality. In the case of C_{60} the
> name is such an individual thing. . . . This is the only little bit of
> individual immortality in the discovery.[2]

When chemists prepare a new substance, they have to name it. There is
an unambiguous but complicated international system of nomenclature
for naming chemical substances. When the stable cage-like C_{60} molecule
was first observed, the discoverers did not establish its 'official' name,
which came later and proved to be prohibitively long and impossible.
They gave it a so-called trivial name, buckminsterfullerene. It is easy to
associate this name with the molecule's structure, and it rolls off one's
tongue easily. It is a beautiful link to design science, represented by R.
Buckminster Fuller, whose United States Pavilion at the Montreal 1967
Exposition served as an inspiration in the search for the structure. The
ending 'ene' of the name signals an important chemical property of the
molecule, viz., its unsaturated character. Shortened versions of the name
exist, such as buckyball, footballene, and soccerene. At one point during
the initial phase of the discovery the names Bucky and Mrs. Bucky were
mentioned for C_{60}, and the second most abundant species, C_{70}. The name
'fullerenes' refers to the whole class of similar molecules.

An important name figured in the first-ever Nobel Prize. Wilhelm
Röntgen (P01) called his mysterious radiation X-rays; the English-speaking
world has used this term ever since, while the rest of the world know them
as Röntgen rays.

The importance of naming was conspicuous in the case of element
106. While the name was being disputed, as no element had ever been
called after a living person, Glenn Seaborg (C51) declared that he would
be willing to exchange his Nobel Prize for the name Seaborgium. Come to

think of it, the number of Nobel laureates increases annually by at least five, usually more, while the number of elements in the periodic table is limited. Seaborg had a special fondness for the periodic table, and with good reason. He was a co-discoverer of plutonium, americium, curium, berkelium, californium, einsteinium, fermium, mendelevium, nobelium, and what finally became seaborgium. Furthermore, he formulated the concept of the actinide elements by analogy to the lanthanides, and this is conspicuously reflected in every periodic table. He considered this to be his most important achievement.[3]

Stanley Prusiner (M97) hypothesized that the infectious agent of scrapie and Creutzfeldt–Jakob disease was a single protein, and he called it prion. The name, which stands for 'proteinaceous infectious particle', was a success and contributed greatly to Prusiner's recognition. The Nobel Assembly of the Karolinska Institute took a risk in awarding the prize because many consider this protein-only mechanism of infection unproven as yet.

George Olah (C94) announced the preparation and observation of stable, long-lived carbonium ions in superacid media as early as in 1962. Then, in 1972, he laid out a general theory of carbon cations.[4] He introduced the names carbocation, referring to all positively charged ions of carbon, carbenium for CH_3^+, and carbonium for CH_5^+. These names have been generally accepted. The name superacid goes back to J. B. Conant of Harvard University who used it as early as 1927. Then R. J. Gillespie defined it as an acid that is stronger than 100% sulfuric acid. One of Olah's superacids got a special name, Magic Acid, when someone in his lab put a Christmas candle in FSO_3H-SbF_5 and it dissolved. A company set up by one of Olah's former students to commercialize his reagents has marketed it under this name.[5]

In 1954, there was a meeting of the Faraday Society, 'The Study of Fast Reactions'. For Manfred Eigen (C67)[6] it was an important event because he was coming up with faster reactions than anybody before him had. Other authors, preceding Eigen in the program, reported measurements in the seconds range. They also called it 'fast reactions'. Then came another researcher, reporting on 'very fast reactions', those in the milliseconds range. The next speaker used the expression 'extremely fast reactions'. Then came Eigen's lecture reporting about reactions below microseconds. It was suggested that they be called 'damn fast reactions' or even 'damn fast reactions indeed'. He later titled his Nobel lecture 'Immeasurably Fast Reactions'.

Gowland Hopkins (M29) had proposed the idea of vitamins back in 1912. Vitamins are part of a balanced diet, and although only needed in small quantities, they are essential. Deficiency in vitamin C, for example, causes scurvy. Albert Szent-Györgyi (M37), while working with Hopkins in Cambridge in 1928, isolated a compound from the adrenal gland and from cabbage, and determined its composition to be $C_6H_8O_6$; he did not, however, know its structure, that is, the connectivity of the atoms in the molecule. Later it turned out to be vitamin C, and when Szent-Györgyi returned to the University of Szeged in southern Hungary, he could isolate it in large quantities from paprika, a common vegetable in Hungary. When he first described the substance in a manuscript submitted to the *Biochemical Journal*, he named it 'godnose'. But the editor of the journal would not let him use the name, and so although it did not appear, it has become one of the best known no-names in science.

Vladimir Engelhardt[7] in Moscow discovered oxidative phosphorylation, which was the aerobic esterification of inorganic phosphate. It was work that could have earned Engelhardt a Nobel Prize. Engelhardt originally called it 'respiratory resynthesis' of ATP (adenosine triphosphate). This discovery showed that ATP was fundamental in bioenergetics, that it was the storage form, the energy currency by which energy was produced. The corresponding enzyme, ATPase, sets the stored chemical energy free. Engelhardt and his wife, Militza Lyubimova, discovered this enzymatic property of myosin and first suggested the name ATPase (rather than adenosine triphosphatase) in a manuscript submitted to *Nature*. Although the editors rejected the name (while accepting the paper[8]), ATPase has become generally accepted and widely used.

Engelhardt was the first director, of what is today the V. A. Engelhardt Institute of Molecular Biology of the Russian Academy of Sciences, when the dictator of Soviet agro-science, T. D. Lysenko, still reigned. The term molecular biology was anathema to the Soviet authorities, so his institute was first named Institute of Physico-Chemical and Radiation Biology of the USSR Academy of Sciences. Participants at the 1961 Moscow International Congress of Biochemistry were also forbidden to use the name.[9] There was to be a section on molecular biology, for which Max Perutz (C62) and Engelhardt were jointly responsible. Perutz had accepted the task on condition that he should have a free hand in compiling the scientific program, and this is indeed what happened. However, when Engelhardt proposed the name molecular biology to the Soviet organizing committee, everybody else opposed it; hence the title of the section became 'biological functions at the molecular level'.[10]

It was Warren Weaver who first used the term 'molecular biology', in 1938[11] in an annual report of the Rockefeller Foundation, whose natural science section began with a 16-page discussion entitled 'Molecular Biology'. The first sentence of the report reads as follows: 'Among the studies to which the Foundation is giving support is a series in a relatively new field, which may be called molecular biology, in which delicate modern techniques are being used to investigate ever more minute details of certain life processes.' The report signified a major shift in emphasis of support by the Rockefeller Foundation, from the previous physics orientation to biological problems, which had begun when Warren Weaver became head of its natural sciences section in 1932. Weaver was a trained mathematician but had been lured away from a mathematics professorial post by the natural sciences directorship at the Rockefeller.[12]

Molecular biology has proved to be a hit although there was opposition to the name from diverse corners. Erwin Chargaff's statement became famous: 'molecular biology is the practice of biochemistry without a license'. It resonates with a later opinion by Kary Mullis (C93), who said that molecular biologists are not very interested in chemistry, 'They wish that life was based on something else.'[13] The British geneticist Conrad H. Waddington suggested the name 'ultrastructural biology' instead of molecular biology,[14] while Salvador Luria (M69) complained that the term evoked irritation in some and confusion in others. There is a notion that in the early 2000s we live in a post-molecular biology era, in that the techniques of molecular biology have penetrated all biology and its program has been fulfilled. The two roots of molecular biology, structural chemistry and genetic information, have merged in the realization of how the one-dimensional genetic information contained in DNA is converted into three-dimensional protein structures.

Luria[15] used the term 'supramolecular biology' for studying the organization of the supramolecular structures of living cells. In the meantime, the term 'supramolecular chemistry' has also been introduced and become popular. It embraces large systems of molecules that are capable of stable existence separately but which are joined together by bonds weaker than the covalent bond. The chemistry prize of Donald J. Cram, Jean-Marie Lehn, and Charles J. Pedersen in 1987 was for this new field. Pedersen produced some neutral organic molecules capable of forming complexes with inorganic salts, and due to their shape he named them 'crown ethers'. With these crown ethers the positive ions of the salts 'could be crowned or uncrowned without physical damage to either just like heads of royalty.'[16]

Donald Cram[17] coined the expression 'host–guest chemistry' and the more specific one, 'container chemistry'. Container chemistry is applied mainly to capsular complexes such as spheraplexes and carceplexes, in which decomplexation is mechanically inhibited. Molecular container is more suggestive of their geometry than the more general term host. Cram stressed that 'good nomenclature elicits images and aids reasoning by analogy, it is the organic chemist's "best friend"'. Cram established a research program of designing, synthesizing, and studying the binding properties of organic compounds containing enforced concave surfaces. The molecules are shaped like saucers, bowls, and vases. They called them *cavitands*, and their complexes *caviplexes*. When two cavitands are attached together at their rims, they compose closed-surface compounds with large enough enforced interiors to imprison guest molecules. The hosts are called *carcerands*, from the Latin *carcer*, prison.

Another highly successful name is 'macromolecule'. Hermann Staudinger had introduced it long before he received the Nobel Prize in Chemistry for 1953, 'for his discoveries in the field of macromolecular chemistry'. 'Polymer' is also popular for the same systems, while 'giant molecule', suggested for molecular weights between 12 000 and 15 000, did not take root. Emil Fischer (C02) came up with this name but he also held that the highest molecular weight for synthetic organic substances would be around 4000. The names macromolecule and polymer signified a definite stand in chemistry during the first half of the twentieth century, when a fierce debate was raging over whether biological substances consisted of colloidal systems or macromolecules.[18] This controversial situation must have contributed to the lateness of Staudinger's Nobel Prize. Another one, to Karl Ziegler and Giulio Natta, followed in 1963 'for their discoveries in the field of chemistry and technology of high polymers'. There was yet one more, in 1974, to Paul J. Flory 'for his fundamental achievements, both theoretical and experimental, in the physical chemistry of the macromolecules'. When *Time* magazine did a survey to find the greatest scientists of the twentieth century, L. H. Baekeland headed the field of chemistry. He was an academic turned technologist and invented the widely used synthetic plastic, Bakelite. In the Nobel Archives he figured once, as a nominator of I. Langmuir (C32) in 1928, but never as nominee.

The quarks are the six fundamental elementary particles, and it is from combinations formed by strong interactions between them that more familiar particles, such as protons and neutrons, are built. Their existence was deduced by physicist Robert Serber of Columbia University, when he concluded that protons and neutrons (with their common name, baryons) and another group of particles were not truly elementary but were made of particles of yet a more fundamental elementarity. Serber discussed the idea with Murray Gell-Mann (P69) and they together came to the conclusion about the fractional charges of these 'new' particles. One story about how the name came about relates that Gell-Mann remarked that the existence of such particles would be a strange *quirk* of nature, which was then transformed to *quark*.[19] However, other stories about the origin of quark are also known.

Gell-Mann and Yuval Ne'eman both did fundamental work in the classification of elementary particles, but only Gell-Mann recognized the power of naming. 'Quarks are elementary particles, building blocks of the atomic nucleus. I am one of the two theorists who predicted their existence and it was I who gave them their name.'[20] He also originated such names as the 'Eight-fold Way' and 'strangeness' in elementary particle physics.

Otto Hahn and Lise Meitner worked for years in Berlin on bombarding uranium with neutrons and studying the radioactive substances that were formed. By the time Hahn and his associate Strassman made a successful experiment and were ready to identify the products in 1938, Lise Meitner had fled from Germany to Sweden. During the Christmas vacation she and her nephew, Otto Frisch, were interpreting Hahn and Strassman's experimental results. Gradually they came to the conclusion that the uranium nucleus could be visualized as a wobbly unstable liquid drop, which would divide itself at the impact of a neutron.[21] It reminded Frisch of the process in which single cells divide into two. As the paper was being prepared for publication, a biologist told Frisch that cell division was called fission. Thus Frisch coined the term 'nuclear fission.'[22] Niels Bohr (P22) did not like this name[23] and he was deeply involved in the field; he and John Wheeler had worked out the detailed theory of nuclear fission.[24] One of Bohr's objections was that there was no corresponding verb to fission. After trying out a few alternatives, though, they accepted fission. Wheeler himself coined some highly successful names, such as 'black hole'.

Publishing

There are over a million researchers in science publishing over a million articles annually in thousands of scientific periodicals. In spite of this tremendous proliferation of information, the important results get delineated from the rest and find their way into a select set of publications. The British *Nature* and the American *Science* are the most visible and most authoritative vehicles for bringing out new results, and they can make or break research careers. The publication of poor science is easily rejected by good periodicals. Solid and reliable, but mediocre papers may have the safest sailing through the peer review process. It is noteworthy how often people find it difficult to have their (eventually) Nobel Prize-winning papers accepted for publication in the most prestigious journals. We will see two examples below, both successful, in which the scientist went to a journal outside his area of research, hoping that it would be easier to have his unorthodox results published. The former editor of *Nature*, John Maddox, thinks that refereeing may be too strict:

> *Nature* could do quite a lot to demystify itself, and it ought to be less rigorous in seeking the approval of referees for everything it publishes. There are many fields of science where, inevitably, the difficulties are so great that people cannot provide the kind of proof that is available in other fields. For example, in neuroscience it is very hard to have that kind of proof. If somebody wanted to write a paper on the meaning of consciousness in *Nature*, chances are that it would not be published because the referees dealing with topics like that are so steeped in the idea that everything must be backed up by experiment and so on.[25]

It was under Maddox's editorship that Kary Mullis' manuscript introducing PCR was rejected by *Nature*. Even after many years, Maddox[25] remembers why it was not accepted. *Nature* requested that Mullis describe an example of applying PCR; Mullis did not comply, and published his discovery elsewhere.[26]

Another example of *Nature* rejecting publication of Nobel Prize-winning research was the report about the successful crystallization of the substance (*Rhodopseudomonas viridis*) that then served to determine the structure of the photosynthetic reaction center in bacteria. Previous attempts at crystallization had failed for many years. The paper was published in another journal,[27] but it was a consolation to Maddox that the full description of the structure determination later appeared in *Nature*.[28]

In 1937, when Hans Krebs (M53) had unraveled the nature of the citric acid cycle, he wanted to report the discovery in *Nature*. However, the editor wrote him that the letter section of the journal was for the moment full and that Dr. Krebs could resubmit his paper if he wished in some months' time, but that as an alternative he might consider publishing elsewhere, which amounted to a polite rejection.[29] Krebs went on to receive the Nobel Prize in 1953 'for his discovery of the citric acid cycle'.[30]

Paul Boyer's (C97)[31] manuscript on his prize-winning binding-change mechanism had been rejected by the *Journal of Biological Chemistry* before he published it in the *Proceedings of the National Academy of Sciences of the USA*. Boyer thinks it is inevitable that unusual results are difficult to accept.

Gerald Edelman (M72)[32] did not try to publish his first big discovery in any journal in his field, choosing instead to go beyond the reach of his peers.[33] He was studying the structure of an immunoglobulin antibody molecule using ultracentrifugation. In the mid-1950s, when Edelman started his research, it was thought that the molecule was a long polypeptide chain. Edelman wanted to correlate the sedimentation coefficient from ultracentrifugation with the number of disulfide bonds he had cleaved. He observed that the antibody molecules slowed down so much in the centrifugal field that it was impossible to explain it by the opening of the chain and an increase in the frictional coefficient. Rodney Porter (M72), a famous British scientist 12 years Edelman's senior with whom Edelman would share the Nobel Prize, had already published a paper about the molecule being one long chain. Thus Edelman was having a hard time with his notion that the immunoglobulin molecule had independent subunits, linked by disulfide bonds. Edelman worked with a complicated experimental system and he used more sophisticated mathematics than people in biology were used to. This was coupled with his being an unknown name in the field at the time: he had not yet earned his PhD, although he was already a medical doctor. After 196 additional control experiments he decided to publish his findings, but instead of choosing a biological journal he sent his letter to the editor of the *Journal of the American Chemical Society*.[33] Apparently the chemists did not think his proposition to be so impossible.

In the annals of hardship in getting unexpected results published, one of the most striking examples is the story of the oscillating reactions. Although the finding of oscillating reactions did not earn their discoverers the Nobel Prize, it could have. According to Ilya Prigogine (C77),[34] it was one of the most important breakthroughs of the twentieth century, as important as the discovery of quarks or black holes. It demonstrated a new type of coherence that extends over macroscopic distances, a striking example of non-equilibrium structures. Prigogine ascribes the difficulties in publishing the discovery of oscillating reactions to the fact that the results seemed to be in contradiction with thermodynamics. It would have been a contradiction, indeed, had oscillations been observed close to equilibrium, but they were not. These long-range correlations appear far from equilibrium.

Boris Belousov (1893–1970), an obscure Russian military chemist in Moscow, made the original discovery. He put together a relatively simple chemical reaction and observed a periodic change in the reagent concentrations by their oscillating color change. His paper about this reaction was rejected repeatedly by Russian chemical journals, on the grounds that it was suggesting something impossible. Nobody bothered to check the validity of his claims, which in fact would have been easy to do. Finally Belousov gave up, but not before he had managed to describe his findings in a very short note in an obscure, unrefereed publication in 1959.[35] Anatol Zhabotinsky,[36] upon graduating from Moscow University, took up this project. He sent his manuscript to a biophysical rather than a chemical journal and succeeded in publishing his results in 1964.[37] At that time chemists believed that the oscillating reactions contradicted the second law of thermodynamics, but the biophysicists were unaware of this. Today, the Belousov–Zhabotinsky reactions have a vast literature.

Dan Shechtman also found it difficult to publish his discovery, the first experimental observation of quasicrystals. These are regular but non-periodic solid-state structures, which display fivefold symmetry. Such structures used to be deemed impossible by classical crystallography. The first rejection was all the more painful because it did not contest his assertion, it simply said that there would be no interest in his observation. When, after two years, Shechtman finally got through, in another journal,[38] thousands of publications followed.

The world learned about Odd Hassel's (C69) prize-winning discovery of conformational equilibrium after the Second World War from a paper in *Nature*.[39] Originally the discovery was published in 1943 in a little-known Norwegian journal.[40] Norway was under German occupation at the time and Hassel was not allowed to publish his paper in English. He refused to publish it in German so it appeared in Norwegian. It was an article often quoted but seldom read, until it was republished in English translation in the same journal on the 25th anniversary of its original appearance, shortly before Hassel's Nobel Prize.

Rita Levi-Montalcini (M86) struggled to publish the first results from her studies of the nervous centers of chick embryos. In 1938 the Italian authorities started introducing laws restricting the activities of Jews and she could not publish in any Italian scientific journal. Somewhat later her paper appeared in an excellent Swiss periodical.[41]

Sherwood Rowland and Mario Molina also met all kinds of difficulties when they wanted to publish their findings about the depletion of the ozone layer. Rowland sent the manuscript to *Nature* and it took about five months to get it published rather than the customary few weeks.[42] In addition to publishing their results, Rowland was looking for additional publicity and they arranged for a press conference at the University of California campus at Irvine. The *Los Angeles Times* published an article about their work, but only in its Orange County edition. The news agency Reuters issued a story, but as it did not identify the authors by name and school there was no follow-up. Rowland and Molina were disappointed and wrote a much longer document. To speed up its publication, they labeled this review as a report of the Atomic Energy Commission (AEC), the body that funded their work. This did not require any refereeing and made it possible to give out copies of the document; they gave out hundreds of them. At the next meeting of the American Chemical Society,

Rowland was one of the participants in the customary press conference. Associated Press picked up his story, which went to 400 newspapers with a total circulation of 100 million. The ball started rolling and CFCs and ozone became a big media feature.

Arthur Kornberg's (M59)[43] first paper on enzymatic synthesis of DNA, his prize-winning research, had difficulties in getting accepted by the *Journal of Biological Chemistry*. Although it was not rejected outright, the editors wanted him to call the DNA polymerase product a poly-deoxyribonucleotide rather than DNA. There were many reviewers and a lot of controversy, but Kornberg would not change the title of the paper and it took the arrival of a new editor at the journal to finally get it accepted.

Prior to Marshall Nirenberg's (M68)[44] bombastic announcement about cracking the genetic code at the Moscow biochemistry meeting, Matthaei and Nirenberg published a note in *Biochemical and Biophysical Research Communications*. It described their assay, showing that ribosomal RNA stimulates amino acids into proteins, but caused no commotion. On the other hand, Nirenberg was reluctant to spend much time on writing papers, which would have taken away precious time from his experiments. He went for almost two years without publishing a single paper. Immediately before he left for Moscow he had written two papers for the *Proceedings of the National Academy of Sciences of the U.S.A.* Since he was not a member of the Academy, Joseph Smadel, the vice-director of the NIH, submitted them. However, when he later asked Smadel to sponsor yet another of his papers in the *Proceedings*, Smadel said, 'Nirenberg, I've done enough for you.'

At that point, Nirenberg turned to Leo Szilard for help. Szilard was a member of the National Academy of Sciences, and he was in Washington so it was convenient for the Washington-based Nirenberg to ask him to sponsor his next papers in the Academy's proceedings. When Nirenberg called him, Szilard invited him down to Dupont Hotel where Szilard lived at that time, on Dupont Circle. Szilard was deeply involved in defense matters and used the hotel lobby as his office. He knew all of the people from the Pentagon and 'official' Washington; they would come to the hotel and Szilard would confer with them. So Nirenberg came and since Szilard was a physicist he asked Nirenberg to explain the work to him. Nirenberg spent the day talking to Szilard in the lobby, explaining what they had done and what the implications were, with people who were passing through constantly interrupting them. The story is vintage Szilard. At the end Szilard said, 'It's too much out of my field. I'm sorry, I can't sponsor it.'

Sidney Altman (C89),[45] who shared his Nobel Prize with Thomas Cech, 'for their discovery of catalytic properties of RNA' found it 'extremely difficult' to get his results published. Various journals just rejected them out of hand, and it took about two years to get the results out. Altman resorted to the same technique as Nirenberg: a senior colleague of his had it refereed and sent it to the *Proceedings of the National Academy of Sciences*.

In her Nobel lecture Rosalyn Yalow (M77) showed, and in the printed version of the lecture reproduced, a letter of rejection of their prize-winning paper by a journal.[46] Berson, Yalow, and their co-workers demonstrated the ubiquitous presence of insulin-binding antibodies in insulin-treated subjects. Their original paper was rejected by *Science* magazine, and was initially also rejected by the *Journal of Clinical Investigation*. The latter's comments included: 'The experts in this field have been particularly emphatic in rejecting your positive statement.'

Günter Blobel (M99) also experienced difficulties in publishing his findings. One of his most important papers was a combination of biophysics and electrophysiology, and demonstrated the existence of a proteinconducting channel. The editors of *Science* 'tortured us because they had terrible reviewers. One of them declared that our paper was of no interest to anybody. Another said that it was all wrong, it was all artifacts. The manuscript went back and forth for three or four months. Finally they said they would publish a thousand words or something like that.'[47] So Blobel sent it to another journal, *Cell*, which accepted it within 24 hours and published it four weeks later.

George Olah (C94)[48] was still in an industrial laboratory, at 35, relatively young and a recent immigrant, when he was invited to give a major lecture at a conference in 1962 at the Brookhaven National Laboratory. Then and there he first announced the preparation and observation of stable, long-lived carbocations. It was at the time of the ongoing debate between two great organic chemists, Herbert Brown (C79) and Saul Winstein, on some mechanistic problems in organic chemistry. Eventually Olah's discovery was to settle the controversy. It was lucky for Olah that there was this big disagreement between two highly visible scientists:

people recall the spectacle of the debates, which took place at large meetings almost as if carefully choreographed. The details are no longer of interest, but the essence was that the two adversaries interpreted the same experimental data in different ways. To be sure, nobody doubted the validity of the experimental data; it was the mechanism of the reaction that was unknown and being contested.[49]

According to the presentation speech at the Nobel ceremony in 1994, on the occasion of Olah's Nobel award, before his discovery it was like seeing the first and last scenes of *Hamlet* without knowing what happened in between.[50] By using his superacids, Olah lent longer life to his carbocations and made their observation possible. This led to the resolution of the big controversy. But back in 1962, Olah's announcement was greeted with near disbelief. Both Winstein and Brown called him aside during the conference and cautioned him that a young chemist should be exceedingly careful when making claims. Each pointed out that most probably Olah was wrong and could not have obtained long-lived carbonium ions. Just in case Olah's method should turn out to be real, however, each of the two men expected him to provide evidence for his respective viewpoint. Resolving the controversy was a stepping stone to Stockholm for Olah, and gave added publicity to his discovery.

Lawrence (W. L.) Bragg (P15) taught his pupils 'to publish their results promptly, lucidly and concisely'. Bragg could write a manuscript in an evening, and have it typed next day ready for submission. Perutz (C62) compared it to Mozart writing the overture to *The Marriage of Figaro* in a single night.[51] Bragg noted with clarity the importance of timely publishing:

> When a major discovery in science is made, there is always a number of people who claim that the idea or the discovery was implicit in some previous work they or others have done; very often they are right. Such still-born children of the scientific brain abound, still-born because their author had not the art, the confidence, the enthusiasm and vitality to express his work in such a form that it had a living impact on the world of science. As I sometimes feel it necessary to remind young research students, we are not writing our papers for consideration only by God and a committee of archangels, but for frail fellow mortals. . . . Unless a paper has an immediate effect, it will almost certainly play no part at all in the progress of science and might as well never have been written.[51]

There is some ruthlessness in the way aging scientific contributions are viewed, and Bragg put it eloquently:

> Papers of the last generation are only of interest to science historians. Here science is unlike the arts, where the value of original thought is often enhanced by time. Science is like a coral reef, alive only on the growing surface. The work of the past is of course the foundation on which further advance has been made, but it is dead, it has been replaced by a more complete understanding.[51]

John Maddox sounds rather pessimistic when he says:

> everything *Nature* publishes will be proved wrong within a measurable period of time, perhaps 10 years, perhaps 20 years, but, in due course, all the stuff we're publishing now will turn out to be technically wrong. New things will be discovered to show that the hypotheses that people believed to be correct in 1998 will no longer be correct.[25]

John Cornforth's (C75) words may serve as a caveat about publications. Because of his deafness he has a particularly sharp eye for the literature. He cannot draw much from meetings or from conversations involving more than two people, so he goes for the printed record and has critical views worth considering:

> I disbelieve in abstracts and reviews automatically. An abstract is what the abstractor thought was important of what an editor and his referees allowed to appear of what the author thought was important! By the time you read either a review or an abstract it has been filtered through too many brains. The best you can do is to go for the original paper, but what I get from a paper is almost never what was intended. So I am worried about the tendency to avoid full publication of results and to rely on databases, while access to the whole of the literature is becoming more difficult because libraries cannot afford full coverage. The quantity of information being produced is stupendous but the quality is nobody's business. The usual 'preliminary' publication is almost all interpretation, not the actual experiments, which are the only things of lasting value. . . . I take a lot of care with my papers and I am content that they will be my memorial. 'What you did, why you did it, and what were your results' is as valid an ideal as when Rutherford enunciated it.[52]

Citations

An important ingredient in having an impact is the publicity of publications, and the number of citations received for them is a telling measure of this publicity. Eugene Garfield, the founder of *Science Citation Index*, started a revolution in evaluating scientists and research venues by considering the number of citations received on their publications. However, although the Nobel laureates had published highly cited papers, there is no direct correlation between citation records and Nobel Prizes. Ad Bax of the National Institutes of Health, doing protein nuclear magnetic resonance spectroscopy, heads the list of the most cited 50 chemists in the period 1981 to 1997. As the list contains only nine Nobel laureates, there are quite a few missing; however, they may be high up on the citation lists of their more narrow specialization. Contrasting the nine Nobel laureates on the list with the top nine non-Nobel authors suggests that the laureates tended to publish fewer papers, yet more highly cited ones, than the top authors on the list.

Oliver Lowry is the most cited scientist of all time with over 250 000 citations for an article[53] in which he gave a methodology for protein measurement. Methodology papers do tend to get more citations than theoretical papers but Lowry's citation record is extraordinary. There is a 'Lowry factor' in the citation literature in that, for more realistic evaluation of all other authors' data, Lowry's parameters are usually removed from the comparisons. Lowry was not a Nobel laureate but had a pivotal role in launching the career of at least one: Daniel Nathans (M78)[54] (cf. p. 124).

The most cited British scientist at the end of the twentieth century, and by a huge margin, was Salvador Moncada at University College London. Moncada is most famous for his nitric oxide discovery, but was not included in the nitric oxide Nobel Prize in 1998 (cf. p. 224).

Citation data are important in the evaluation of impact and have the advantage of being quantitative and objective. It is, though, only one of several ingredients that may be considered in the evaluation of a scientist's performance. For the Nobel Prize it is more of a curiosity than a decisive factor.

Citation data must be interpreted with care, due to the existence of two interesting phenomena. The first is when a publication does not have an immediate impact, for whatever reason. It may be that the discovery is ahead of its time, or that the report was communicated in the wrong journal. For James Black's (M88) prize-winning discovery of the beta blockers, it was decisive for him to learn about the α and β receptors that mediate the physiological effects of adrenaline. Although the idea that drugs acted on receptors was an old one, it had never been used to explain physiological phenomena. Black read about it in a book chapter by Raymond Ahlquist, who had had great difficulties in getting his findings published, and when he finally did, had his paper ignored for another ten years.

Garfield calls the other phenomenon 'obliteration by incorporation'. Some seminal discoveries become so much a part of the culture of science that they are usually mentioned without reference to the original source. Examples include the periodic table of the elements by Dmitri Mendeleev and the double helix structure of DNA by Watson (M62) and Crick (M62). No wonder the respective original publications are not cited in record numbers.

11

Is there life after the Nobel Prize?

The Nobel Prize changes the lives of the laureates. They have control over some changes, and hardly any control over others. A case in point is Frederick Robbins (M54).[1] He was a member of the team that succeeded in growing the poliomyelitis virus in tissue cultures. He was also a pediatrician, but from the day of the Nobel announcement he never received another request to see a patient in consultation. He was a professor at a university so his livelihood did not depend on it, but it is symptomatic of how high a pedestal is instantly created for the Nobel laureate as a consequence of the prize. As another winner stated succinctly: 'Winning the Nobel Prize is hazardous to your health.'[2] Alfred Gilman (M94),[3] on the other hand, compared the impact of the Nobel Prize on him to receiving a lifetime's supply of Prozac injections.

Nobel laureates often like to maintain that their life and activities have not changed, apart from a temporary disruption in their routine, although the reality may be different. Nonetheless, those laureates whose scientific life came to an end after the prize made this choice themselves. Abraham Pais wrote of Isidor Rabi: 'The Nobel Prize had diminished his ardent pursuit of pure science.'[4] Then Pais quotes someone else: 'Unless you are very competitive you aren't likely to function with the same vigor afterward. It's like the lady from Boston who said, "Why should I travel when I'm already there?"'[5] It is not at all uncommon, especially outside the United States and Great Britain, for successful scientists, regardless of whether they win the Nobel Prize or not, to gradually shift towards public life and away from the laboratory. One of Owen Chamberlain's (P59) colleagues predicted that the 39-year-old fresh Nobel laureate had already done his last experiment, reflecting the widely held view that outside pressures would prevent him from carrying on as a researcher. Chamberlain promised himself that he would not succumb to such pressures, and he did succeed in continuing as an experimental physicist.[6] Melvin Schwartz (P88), on the other hand, was lured back to science by winning the prize. He had left Stanford University and was running a successful business, only to return to academia in the wake of his award.[7]

The Nobel Prize causes considerable commotion, even in institutions for which it is not too extraordinary. Rudolf Mössbauer (P61) was awarded the prize after having just moved from Germany to Cal Tech. On the morning of the announcement, Cal Tech disconnected his telephone so he would not be bothered, and two sheriff's cars appeared in front of his house to prevent reporters from getting to him.[8] Edward Lewis (M95), also of Cal Tech, donned a fake beard when his Nobel Prize was announced. Although ostensibly for disguise, I suspect he was wearing it as a parody.

The impact of the Nobel Prize on the individual also depends on his circumstances. An extreme case was when Robert Bárány[9] was awarded the physiology or medicine prize in 1914 'for his work on the physiology and pathology of the vestibular apparatus'. The telegram notifying him about the award found him in a prisoner-of-war camp in southern Russia, in Turkmenistan, just north of Afghanistan. He had been captured, while serving as a medical officer in the Austro-Hungarian Army, during the early months of the First World War. His Nobel Prize called attention to the fact that he was not being afforded the conditions to which he was entitled by his academic degree. Prince Carl, the head of the Swedish Red Cross, had initiated an agreement whereby POWs with an academic degree should be given work at universities. Thus Bárány was transferred to a university somewhere in eastern Russia, and from then on he treated patients and did research. One of his patients, a grand duke, noticed his limp, which was a consequence of childhood bone tuberculosis. He advised Bárány to get a stick and start limping even more strongly, and this qualified him to be a medical victim of the war. In June 1916 Bárány was exchanged at the Swedish–Russian border (Finland was part of Sweden at that time). He went home to Vienna and returned to Stockholm in September 1916 to receive the Nobel Prize and deliver his lecture. When jealousy and anti-Semitism in Vienna made his life unbearable, Bárány returned to Sweden and spent the rest of his life as principal and professor of an otological institute in Uppsala.

Referring to the benefits of his Nobel Prize, Jerome Karle (C85) mentioned 'the opportunity to have contact to an unprecedented degree with young people who look forward to careers in science and other intellectual and artistic pursuits'.[10] Karle works in the Naval Research Laboratory in Washington where access to him is restricted, and the prize may have made him more accessible. On a lighter note, Chamberlain (P59) remembers the reaction to the announcement of the Nobel Prize by one of his children. She said, 'Does that mean that one of Dad's experiments worked?'[11] The Nobel award freed Herbert Brown (C79) from taking out the household garbage, at least for a while, if one can believe the newspaper cartoons.[12]

There are Nobel laureates who get increasingly involved in 'outside' activities, such as education, collecting art, writing poetry, or standing up for political causes, while for others the prize helped them to continue scientific work long after retirement age.

A remarkable exception is Frederick Sanger (C58 and C80). He continued his research without any disruption after his first Nobel Prize, and considers the main benefit from it to have been a steady research position at the MRC.[13] After his second Nobel Prize he continued his laboratory experiments until the very day when he reached retirement, at age 65 in 1983. On the eve of his retirement Sanger was sitting at the laboratory bench carrying out an experiment. The next morning his laboratory was empty.[14] He has been into gardening ever since, and makes no statements about politics, art, or literature.[13]

Proper employment to the Nobel laureate's liking is not always a trivial matter. Daniel Bovet (M57), who had previously worked at the National Health Institute (NHI) in Rome, in 1963 applied for a professorship in the framework of the complex Italian system of competition, making his intentions known at the last minute. At the time of his application the only available chair was at a remote university in northern Sardinia, but he was determined to leave the NHI where conditions had become unbearable. The professorial appointments are usually distributed at an early stage in the intricate committee arrangements, yet a Nobel laureate could not be denied a professorship. It was clear to everyone that the committee would have preferred it if no Nobel laureate had applied. Although his opponents tried to reject him on the basis of his lack of teaching experience, he was given the job. It took him years to get back to Rome, an experience about which Bovet was bitter. He kept a journal about his ordeal and deposited it in the Pasteur Institute in Paris; it will become available for research only in 2042.[15]

Socially

Laureates seldom admit the beneficial changes in their social position following the Nobel Prize although they are usually proud of their personalized parking spots on busy American campuses. As a refreshing exception, Herbert Hauptman (C85) noted the change in the social status of he and his wife in their community, and especially in that of his wife who has been sought out by society ladies more often since the Nobel Prize than before.

John Vane (M82) appreciated the opportunities the Nobel Prize gave him to have a higher political profile if he had so wished, although he did not utilize the opportunity. He turned down invitations to lecture on topics about which he was not well informed, and he does not even attend the traditional triennial meetings for Nobel laureates in Lindau, Germany, because he does not want to get involved in the politics of Nobel Prize winners. He welcomed the knighthood, though, which he thinks counted more towards changing his life than the Nobel Prize. He lives in London, where the knighthood enables him to get seats in theaters, and so on. He says: 'It is more obvious because now, when I am on the phone I am able to say, "This is Sir John speaking."'[16] Vane's wife thinks that their lives went into a different orbit.

Kary Mullis (C93) has no qualms about letting people know that he is a Nobel laureate: 'Nobody in the world doesn't understand the weight of the Nobel Prize. Once you have it, there is not a single office in the world that you can't go into. If I call them and say, I would like to talk to you about something, and I'm so-and-so, the Nobel laureate, they'll see me at least once. It opens every door.'[17]

César Milstein (M84)[14] did not enjoy publicity, but he became a popular figure back in his native Argentina where Nobel Prizes are scarce. When the dictatorship was over in Argentina, people became interested in the recent history of their country. Milstein got into the limelight in view of his Nobel Prize, and when the facts about his letter of resignation became known (p. 113). Many considered him a symbol of what the country has

Ne'eman, Yuval (1925–), Stockholm, 2000

Oliphant, Mark (1901–2000), Canberra, 1999

Osawa, Eiji (1935–), Toyohashi, Japan, 1994

Pitzer, Kenneth (1914–97), Berkeley, California, 1996

Ringertz, Nils (1932–), Stockholm, 1999

Shechtman, Dan (1941–), Balatonfüred, Hungary, 1995

Shoenberg, David (1911–), Cambridge, 2000

Stork, Gilbert (1923–), New York, 1999

Szent-Györgyi, Andrew (1924–), Budapest, 1999

Teller, Edward (1908–), Stanford, California, 1996

Weissmann, Charles (1931–), London, 2000

Westheimer, Frank (1912–), Squam Lake, New Hampshire, 1995

Whetten, Robert (1959–), Erice, Italy, 1995

Zhabotinsky, Anatol (1938–), Waltham, Massachusetts, 1995

Most of these encounters have been recorded, and the often in-depth interviews are being published in my *Candid Science*[1] book series, along with interviews with other great scientists.

I have also used the video interviews with Eugene Wigner (P63), Emilio Segrè (P59), and Malvin Calvin (C61) by Clarence E. Larson (1909–99) in the framework of his and his wife, Jane's, 'Pioneers of Science and Technology' project. My wife and I are grateful to Jane Larson who graciously asked us to take over and use their original videotape collection.

For their kind help at several levels, I am grateful to my Swedish friends and colleagues:

Professor of Physics Anders Bárány, secretary of the Nobel Committee for Physics and senior curator of the Nobel Museum (NM)

Professor Emeritus of Physics Ingmar Bergström, former director of the Manne Siegbahn Institute of Physics, Stockholm University

The late Professor of Organic Chemistry of Lund University, Lennart Eberson, former chairman of the Nobel Committee for Chemistry

The late Professor Emeritus of Biochemistry of Stockholm University, Lars Ernster, former member of the Nobel Committee for Chemistry

Professor Emeritus of Tumor Biology of the Karolinska Institute, George Klein, former chairman of the Nobel Assembly of the Karolinska Institute

Professor Emeritus of Physiological and Medical Chemistry of Uppsala University, Torvard C. Laurent, chairman of the Board of Trustees of the Nobel Foundation and former president of the Royal Swedish Academy of Sciences

Professor Emeritus of Cell and Molecular Biology at the Karolinska Institute, Nils Ringertz, secretary of the Nobel Committee for Physiology or Medicine and project director of the Nobel e-Museum (NeM); and the Archives at the Center for History of Science of the Royal Swedish Academy of Sciences (Nobel Archives) and its assistant director, Karl Grandin, and the Archives of the Karolinska Institute (through a visit by Magdolna Hargittai).

❦❦❦

I appreciate the further suggestions, literature, illustrative material, and other help I have received from the following: Aldo Domenicano (L'Aquila, Italy); John Emsley (Cambridge); Edit Ernster (Stockholm); Walter Gratzer (London); Balázs Hargittai (Loretto, Pennsylvania); Eszter Hargittai (Princeton, New Jersey); Richard Henderson (Cambridge); William B. Jensen and the Oesper Collection in the History of Chemistry, University of Cincinnati (Cincinnati, Ohio); Eva Klein (Stockholm); Kirsty Knott and the Archives of the MRC Laboratory of Molecular Biology (Cambridge); Alan L. Mackay (London); Ramah McKay and the Archives of Cold Spring Harbor Laboratory (Cold Spring Harbor, New York); Yuval Ne'eman (Tel Aviv); Gábor Palló (Budapest); Simon Phillips (Leeds); Stanley Prusiner (San Francisco); David Shoenberg (Cambridge); Peter Tallack (London); and Lev Vilkov (Moscow). I thank István Fábri and Judit Szücs for technical assistance.

I express my gratitude to the following for the critical reading of the manuscript and helpful comments and criticism: Anders Bárány, Stockholm; Walter Gratzer, London; Keith Laidler, Ottawa; Torvard Laurent, Stockholm; Alan Mackay, London; and Gábor Palló, Budapest.

Notes

Preface

1. Rejtö, J. (Howard, P.), *The Fourteen-carat Car*. Nova, Budapest, 1940 (in Hungarian).

2. C56 refers to the Nobel Prize in Chemistry for 1956; similarly, P refers to physics, and M to physiology or medicine.

3. Harry Kroto and Richard Smalley in 1996, John Pople in 1998, and Ahmed Zewail in 1999.

4. Watson, J. D., *The Double Helix: A Personal Account of the Discovery of the Structure of DNA*. New American Library, New York, 1968, p. ix.

5. Weart, S. R.; Szilard, G. W. (eds.), *Leo Szilard: His Version of the Facts, Selected Recollections and Correspondence*. MIT Press, Cambridge, Massachusetts, 1978, p. xii.

6. Kornberg, A., interview in Hargittai, I., *Candid Science II: Conversations with Famous Biomedical Scientists*. Imperial College Press, London, 2002, pp. 50–71.

7. Hideki Shirakawa, private communication, 2002.

8. Hargittai, I., *Candid Science: Conversations with Famous Chemists*, Imperial College Press, London, 2000; I. Hargittai, *Candid Science II: Conversations with Famous Biomedical Scientists*, Imperial College Press, London, 2002; I. Hargittai, *Candid Science III: More Conversations with Famous Chemists*, Imperial College Press, 2003; M. Hargittai, I. Hargittai, *Candid Science IV: Conversations with Famous Physicists*, Imperial College Press, London (in preparation).

9. S. Brenner, Banquet Speech, December 10, 2002, Stockholm (amended by Dr. Brenner on January 24, 2003, and quoted with his permission).

1 The Nobel Prize and Sweden

1. Bergström, I., private communication, Stockholm, 8 October 2000.

2. Ernster, L., interview in Hargittai, I., *Candid Science II: Conversations with Famous Biomedical Scientists*. Imperial College Press, London, 2002, pp. 376–95.

3. The examples include Georg von Békésy (M61) 'for his discoveries of the physical mechanism of stimulation within the cochlea', Max Delbrück (M69) for his studies of virus genetics, Allan Cormack (M79) and Godfrey Hounsfield (M79), 'for the development of

computer assisted tomography', and Rosalyn Yalow (M78) 'for the development of radioimmunoassays of peptide hormones'. Here and elsewhere the citations for the Nobel Prizes are quoted from the *Nobel Foundation Directory*.

4. Conversation with Anders Bárány in Stockholm, 2000, unpublished records.

5. *Nobel: The Man & His Prizes*. Third Edition. Edited by the Nobel Foundation and W. Odelberg. American Elsevier, New York, 1972, p. x.

6. *Nobel: The Man & His Prizes*. Third Edition. Edited by the Nobel Foundation and W. Odelberg. American Elsevier, New York, 1972, p. 65.

7. Sohlman, R., in *Nobel: The Man & His Prizes*. Third Edition. Edited by the Nobel Foundation and W. Odelberg. American Elsevier, New York, 1972, pp. 15–72, 46.

8. Polanyi, J., Stockholm City Hall Speech, 10 December 1986, reproduced in *Chem. Intell.*, 1996, **2**(4), 13.

9. Klein, G., interview in Hargittai, I., *Candid Science II: Conversations with Famous Biomedical Scientists*. Imperial College Press, London, 2002, pp. 416–41.

10. Pais, A., *Einstein Lived Here*. Oxford University Press, New York, 1994.

11. *Science*, 1994, **263**, 923.

12. Ferry, G., *Dorothy Hodgkin: A Life*. Granta Books, London, 1998, p. 293.

13. Anderson, P. W., interview in Hargittai, I., *Chem. Intell.*, 2000, **6**(3), 26–32.

14. *Nobel: The Man & His Prizes*. Third Edition. Edited by the Nobel Foundation and W. Odelberg. American Elsevier, New York, 1972, pp. 224–5.

15. Here and hereafter we often use the Academy or the Science Academy to abbreviate the Royal Swedish Academy of Sciences. The Swedish Academy is something different, the organization of 18 Swedish writers that is responsible for the Nobel Prize in Literature.

16. Djerassi, C., interview in Hargittai, I., *Candid Science: Conversations with Famous Chemists*, Imperial College Press, London, 2000, pp. 72–91.

17. Djerassi, C., *Cantor's Dilemma*. Doubleday, New York, 1989.

18. Mullis, K. B., interview in Hargittai, I., *Candid Science II: Conversations with Famous Biomedical Scientists*. Imperial College Press, London, 2002, pp. 182–95.

19. Levi-Montalcini, R., *In Praise of Imperfection: My Life and Work.* Basic Books, New York, 1988, p. 201.

20. They are available as a handsome booklet, published by the Nobel Foundation.

21. In the first half of the twentieth century the Nobel Prize could be awarded posthumously if the person died after 31 January, provided that at least one nomination had arrived for him. Nowadays the Nobel Prize can be given posthumously only if the person dies after the announcements have been made in October. Private communication from Anders Bárány, 2001.

22. Conseil Européen pour la Recherce Nucléaire.

23. Eberson, L., *Chem. Intell.*, 2000, 6(3), 44–9.

24. 1921, 1922, 1924, 1927, 1928, 1929, 1938, and 1940.

25. This is hearsay only as no one would go on record regarding such co-ordination in the nominations.

26. M. Hargittai's conversation with Gösta Ekspong in Stockholm, 2001, unpublished records.

27. Ernster, L., *Biochemical Society Transactions*, 1994, 22, 253–65, pp. 253–4.

28. David Keilin never received the Nobel Prize (see p. 230).

29. Hargittai, I., *Chem. Intell.*, 1999, 5(3), 61–4.

30. Stock, J. T., *Chem. Intell.*, 1997, 3(4), 22–4; 1999, 5(2), 43–5 and 48; 1999, 5(4), 35–8.

31. Glaser, J., *Chem. Intell.*, 1999, 5(4), 4.

32. Crawford, E., *The Beginnings of the Nobel Institution: The Science Prizes, 1901–1915.* Cambridge University Press and Editions de la Maison des Sciences de l'Homme, Paris, 1984, pp. 109–49.

33. Crawford, E., *Historical Studies in the Physical Sciences*, 2000, 31(1), 37–53, p. 40.

34. The Soviet secret police.

35. Rowland, F. S., interview in Hargittai, I., *Candid Science: Conversations with Famous Chemists.* Imperial College Press, London, 2000, pp. 448–65.

36. Nye, M. J., *Before Big Science: The Pursuit of Modern Chemistry and Physics, 1800–1940.* Harvard University Press, Cambridge, Massachusetts, 1996, p. 198.

37. A. Westgren, letter of nomination, 30 January 1943, Nobel Archives.

38. H. von Euler, letter of nomination, 30 January 1943, Nobel Archives.

39. These can be contrasted with, for instance, James Franck's (P25) nominating letters, repeated many times over, in which he jointly proposed Hahn and Meitner in physics. He used forceful, convincing words about the importance of their discovery and the weight of their contributions. See, for example, the letter of nomination by James Franck (P25), 9 January 1943, Nobel Archives.

40. A. Westgren, letter of nomination, 31 January 1944, Nobel Archives.

41. Including one by H. von Euler, 31 January 1944, Nobel Archives.

42. Crawford, E.; Sime, R. L.; Walker, M., *Physics Today*, September 1997, 26–32, p. 30.

43. Westgren, A., in *Nobel Lectures: Chemistry 1942–1962*. World Scientific, Singapore, 1999, p. 73.

44. A. Einstein, telegram of nomination, 19 January 1945, Nobel Archives.

45. Sending a telegram might not be in accordance with the present secrecy requirements for nominations, which now contain the following warning: 'Fax and electronic mail must not be used.'

46. Pauli was at that time at Princeton and upon receiving the invitation to Stockholm for the forthcoming Nobel ceremonies, he replied with the following telegram on 21 November: 'extremely sorry unable attend nobel prize celebration stop becoming american citizen but unfortunately not before december 10th and trip to stockholm would considerably delay procedure stop could I attend next years ceremony like others of my colleagues will do wolfgang pauli.' Nobel Archives.

47. Letters of nomination, H. von Euler, 31 January 1946; T. Svedberg (C26), 30 January 1946; and A. Tiselius (C48), 30 January 1946; Nobel Archives.

48. The web address of the Nobel e-Museum is http://www.nobel.se.

49. Ringertz, N., *Chem. Intell.*, 2000, 6(2), 22–4.

50. *The New York Times*, 12 April 1945.

51. Heilbron, J. L.; Seidel, R. W., *Lawrence and His Laboratory: A History of the Lawrence Berkeley Laboratory*. Vol. 1. University of California Press, Berkeley, 1989, p. 492.

52. I. I. Rabi, letter, 28 February 1945, Nobel Archives.

53. George R. Harrison, letter, 15 December 1943, Nobel Archives.

54. Archibald V. Hill, letter, 29 December 1941, Nobel Archives.

55. Madame Florence, letter, December 1944, Nobel Archives.

56. Zuckerman, H., *Theoret. Medicine*, 1992, 13, 217–31.

57. There is a brochure, 'Anna-Greta and Holger Crafoord Fund', of the Royal Swedish Academy of Sciences.

58. Rubin, M. B., *Chem. Intell.*, 1997, 3(3), 44–9.

59. The web address of the Lasker Foundation is http://www.laskerfoundation.org.

60. 1963, 1967, 1971, 1978, and 1983.

61. Abrahams, M., *The Best of Annals of Improbable Research*. W. H. Freeman, New York, 1998, p. 31.

62. *Chemistry & Industry*, 19 October 1998, p. 828.

63. Abrahams, M., *The Best of Annals of Improbable Research*. W. H. Freeman, New York, 1998, p. 35.

64. Gigg, R., *Chemistry & Industry*, 7 December 1998, p. 946.

2 The Nobel Prize and national politics

1. Tremblay, J.-F., *C&EN*, 18 January 1999, 103–4.

2. Vainshtein, B. K., *Current Science*, 1997, 72, 455–6.

3. Hubel, D. H., in *Nobel Lectures: Physiology or Medicine 1981–1990*. Editor-in-charge T. Frängsmyr, editor J. Lindsten. World Scientific, Singapore, 1993, p. 21.

4. Klug, A., interview in Hargittai, I., *Candid Science II: Conversations with Famous Biomedical Scientists*. Imperial College Press, London, 2002, pp. 306–29.

5. Stevenson, R., *Chemistry in Britain*, June 1996, 33–5, p. 33.

6. Palló, G., unpublished research report, Budapest, private communication, 2000.

7. Emsley, J., *Chemistry in Britain*, November 1997, 43–5.

8. Pople, J. A., interview in Hargittai, I., *Candid Science: Conversations with Famous Chemists*. Imperial College Press, London, 2000, pp. 178–89.

9. Letter by Ernest Walton, 20 November 1951, Nobel Archives.

10. Yet another example refers to an omission from the Nobel Prize. A headline in the 15 October 1998 issue of the *Daily Telegraph* (London) read, 'British scientist "was robbed by US propaganda".' It referred to the omission of Salvador Moncada from the physiology or medicine prize for the nitric oxide discovery (see, also, p. 244).

11. *The Economist*, 27 February 1982, quoted in Mott, N., *A Life in Science*, Taylor and Francis, London, 1986.

12. Ochoa, S., in *Nobel Lectures: Physiology or Medicine 1942–1962*. World Scientific, Singapore, 1999, pp. 663–4.

13. Elguero, J., interview in Hargittai, I., *Chem. Intell.*, 1997, 3(3), 61–4.

14. *Ashah Evening News*, 10 January 1998.

15. Letter by H. Yukawa to A. Westgren, 21 November 1949, Nobel Archives.

16. Bartholomew, J. R., *Osiris*, 1998, 13, 238–84.

17. Bartholomew, J. R., *Osiris*, 1998, 13, 238–84, p. 252.

18. Shigenori Togo, later Japanese minister of foreign affairs.

19. Bartholomew, J. R., *Osiris*, 1998, 13, 238–84, p. 277.

20. Maddox, J., interview in Hargittai, I., *Chem. Intell.*, 1999, 5(2), 53–5, p. 55.

21. This was the second time the F. A. Cotton award was given. (F. A. Cotton won the first one, in 1995.)

22. In Hungarian, as in Japanese, the surname comes first.

23. Hevesy, de G., *Adventures in Radioisotope Research*, 1962, 1, 27.

24. Medawar, P., *The Threat and the Glory: Reflections on Science and Scientists*, Oxford University Press, 1991, p. 44, in a review of Gwyn Macfarlane's book, *Howard Florey: The Making of a Scientist*, Oxford University Press, 1979.

25. Palló, G., *Hevesy György*. Akadémiai Kiadó, Budapest, 1997 (in Hungarian).

26. Kornberg, A., interview in Hargittai, I., *Candid Science II: Conversations with Famous Biomedical Scientists*. Imperial College Press, London, 2001, pp. 50–71.

27. Feldman, B., *The Nobel Prize: A History of Genius, Controversy, and Prestige*. Arcade Publishing, New York, 2000, pp. 407–10.

28. Fridman, S. A., *Jews—Nobel Laureates: A Short Biographical Dictionary*. Dograph, Moscow, 2000 (in Russian).

29. Klein, G., interview in Hargittai, I., *Candid Science II: Conversations with Famous Biomedical Scientists*. Imperial College Press, London, 2002, pp. 416–41.

30. Lederman, L., interview in Hargittai, I.; Hargittai, M., *Chem. Intell.*, 1998, 4(4), 20–9.

31. Jacob, F., interview in Hargittai, I., *Candid Science II: Conversations with Famous Biomedical Scientists*. Imperial College Press, London, 2002, pp. 84–97.

32. *News Release*, California Institute of Technology, 5 August 1998.

33. R. Robinson, letter of nomination, 26 January 1938, Nobel Archives.

34. Letter of R. Kuhn to H. von Euler, 10 November 1939, Nobel Archives.

35. Letter of R. Kuhn to the Royal Swedish Academy of Sciences, 19 October 1948, Nobel Archives.

36. Letter of A. Butenandt to H. von Euler, 11 November 1939, Nobel Archives.

37. Letter of A. Butenandt to the Royal Swedish Academy of Sciences, 10 November 1948, Nobel Archives.

38. Perutz, M. F., *I Wish I'd Made You Angry Earlier: Essays on Science, Scientists, and Humanity*. Oxford University Press, 1998, p. 154.

39. Svartz, N., in *Nobel Lectures: Physiology or Medicine 1922–1941*. World Scientific, Singapore, 1999, p. 489.

40. A detailed account of the background of Hitler's ban and the following events are given in Crawford, E., 'German scientists and Hitler's vendetta against the Nobel Prizes', *Historical Studies in the Physical Sciences*, 2000, 31(1), 37–53.

41. Müller-Hill, B., *Murderous Science: Elimination by Scientific Selection of Jews, Gipsies, and Others in Germany, 1933–1945*. Cold Spring Harbor Laboratory Press, New York, 1998. Original German edition, *Tödliche Wissenschaft*, Rowohlt Taschenbuch Verlag, Reinbek, 1984.

42. Otto Hahn (C44) was one of those who served in a gas unit, while Max Born (P54) refused to do so. See Rose, P. L., *Heisenberg and the Nazi Atomic Bomb Project: A Study in German Culture*. University of California Press, Berkeley, 1998, p. 268.

43. Crawford, E., *Nationalism and Internationalism in Science, 1880–1939*. Cambridge University Press, 1992, pp. 67, 74.

44. Elena Bonner happened to be in Western Europe at the time of the Nobel award and could attend the Oslo ceremony. I was visiting Oslo University then, and was very impressed by the torchlight demonstration in downtown Oslo, marking Sakharov's Nobel Peace Prize.

45. Blokh, A., *Poisk*, Nos. 31–2, 25 July to 7 August 1998, p. 12 (in Russian).

46. Heilbron, J. L.; Seidel, R. W., *Lawrence and His Laboratory: A History of the Lawrence Berkeley Laboratory*. Vol. 1. University of California Press, Berkeley, 1989, p. 491.

47. Blokh, A., *Izvestiya*, 3 February 2000 (in Russian).

48. Kapitsa's letter to his wife, Anna, from Leningrad to Cambridge, 4 December 1934, as quoted in Boag, J. W.; Rubinin, P. E.; Shoenberg, D., *Kapitza in Cambridge and Moscow: Life and Letters of a Russian Physicist*. North-Holland, Amsterdam, 1990, p. 213. Note a slight difference in the transliteration, Kapitsa/Kapitza.

49. N. R. Dhar, letter of nomination, 27 October 1938, Nobel Archives.

50. Oseen, C. W., in *Nobel Lectures: Physics 1922–1941*. World Scientific, Singapore, 1998, p. 137.

51. Palló, G., unpublished research report, private communication, Budapest, 2000. By the time the report was completed, both the official who had initiated the project and his office had disappeared in the wake of the general elections that followed.

52. M. Hargittai's conversation with Gösta Ekspong in Stockholm, 2001, unpublished records.

3 Who wins Nobel Prizes

1. Garfield, E., interview in Hargittai, I., *Chem. Intell.*, 1999, 5(4), 26–31.

2. Garfield, E., *Current Contents*, 1991, No. 11 (March 18), p. 8.

3. Walker, J. E., interview in Hargittai, I., *Candid Science III: More Conversations with Famous Chemists*. Imperial College Press, London, 2003, pp. 280–91.

4. Dubos, R., *Louis Pasteur: Free Lance of Science*. De Capo Press, New York, 1986, p. 368.

5. Most Valuable Player.

6. Mullis, K., *Dancing Naked in the Mind Field*. Pantheon Books, New York, 1998.

7(a). Mallik, D. C. V., *Notes Rec. R. Soc. Lond.*, 2000, 54(1), 67–83.

7(b). Singh, R; Riess, F., *Notes Rec. R. Soc. Lond.*, 2001, 55(2), 267–83.

8. Conversation with Anders Bárány, Stockholm, 2000, unpublished records.

9. Einstein, A., in *Nobel Lectures: Physics 1901–1921*. World Scientific, Singapore, 1998, pp. 482–90.

10. This is from a letter by W. Pauli to O. Klein, quoted by Pais, A., *The Genius of Science: A Portrait Gallery*, Oxford University Press, 2000, p. 138. Klein assumed the professorship in January 1931.

11. Campbell, J., *Rutherford: Scientist Supreme*. AAS Publications, Christchurch, New Zealand, 1999, p. 479.

12. Klein, O., *Nobel Lectures: Physics 1963–1970*. World Scientific, Singapore, 1998, p. 214.

13. Ernster, L., interview in Hargittai, I., *Candid Science II: Conversations with Famous Biomedical Scientists*. Imperial College Press, London, 2002, pp. 376–95.

14. Ernster, L., in *Nobel Lectures: Chemistry 1971–1980*. Editor-in-charge T. Frängsmyr, editor S. Forsén. World Scientific, Singapore, 1993, p. 289.

15. Bernal, J. D., *Biographical Memoirs of Fellows of the Royal Society*, 1963 (Willian Astbury), November, **9**, 1–35, pp. 25–6.

16. Bragg, W. L.; Kendrew, J. C.; Perutz, M. F., *Proc. R. Soc.*, 1950, 203A, 321–57.

17. Bernal, J. D., in *Structural Chemistry and Molecular Biology*. Edited by A. Rich and N. Davidson. W. H. Freeman, San Francisco and London, 1968, pp. 370–9.

18. Perutz, M., *Structural Biology*, 1994, **1**, 667–71.

19. Müller-Hill, B., *The* lac *Operon: A Short History of a Genetic Paradigm*. Walter de Gruyter, Berlin and New York, 1996. This book is a unique combination of modern science, science history, and personal anecdotes.

20. Sverdlov, E. D.; Monastyrskaya, G. S.; Chestukhin, A. V.; Budowsky, E. I., *FEBS Letters*, 1973, **33**, 15–17.

21. *Yuri A. Ovchinnikov: Life and Scientific Activity*. Responsible editor V. T. Ivanov. Nauka, Moscow, 1991 (in Russian).

22. Ruffini, R., in *The Eighth Marcel Grossmann Meeting on Recent Development in Theoretical and Experimental General Relativity, Gravitation, and Relativistic Field Theories*. Edited by T. Piran. World Scientific, Singapore, 1999, pp. xv–xxix. Gribbin, J.; Gribbin M., *Richard Feynman: A Life in Science*. A Dutton Book, New York, 1997, pp. 198–218.

23. Waller, I., *Nobel Lectures: Physics 1963–1970*. World Scientific, Singapore, 1998, p. 297.

24. Conversation with Yuval Ne'eman in Stockholm, 2000, unpublished records.

25. Johnson, G., *Strange Beauty: Murray Gell-Mann and the Revolution in XXth Century Physics*. Knopf, New York, 1999.

26. Glashow, S. with Bova, B., *Interactions: A Journey Through the Mind of a Particle Physicist and the Matter of this World*. Warner Books, New York, 1988, p. 143.

27. Iley, A. J. G., *Physics Today*, September 1996, 44–9, p. 46.

28. Stork, G., interview in Hargittai, I., *Candid Science III: More Conversations with Famous Chemists*. Imperial College Press, London, 2003, pp. 108–19.

29. Chargaff, E., *Experientia*, 1950, **6**, 201–9.

30. A related observation by Chargaff (see preceding note) was that the proportions of the four bases of a DNA molecule were characteristic of the species from which it was taken. This observation was not utilized in the double helix discovery but had enormous importance in understanding the role of DNA as the hereditary substance.

31. Watson, J. D., *The Double Helix: A Personal Account of the Discovery of the Structure of DNA*. New American Library, New York, 1968, p. 86.

32. Watson, J. D.; Hopkins, N. H.; Roberts, J. W.; Steitz, J. A.; Weiner, A. M., *The Molecular Biology of the Gene*. Benjamin/Cummings, Menlo Park, California, 1987. This is the latest (fourth) edition. The first three editions were authored by J. D. Watson alone (the first appeared in 1965, published by W. A. Benjamin, New York).

33. Watson, J. D.; Crick, F. H. C., *Nature*, 1953, **171**, 737–8.

34. Chargaff, E., *Heraclitean Fire: Sketches from a Life before Nature*. Rockefeller University Press, New York, 1978, p. 93.

35. Chargaff, E., *Heraclitean Fire: Sketches from a Life before Nature*. Rockefeller University Press, New York, 1978, p. 94.

36. Chargaff, E., *Voices in the Labyrinth: Nature, Man and Science*. Seaburg Press, New York, 1977, p. 24.

37. Chargaff, E., *Heraclitean Fire: Sketches from a Life before Nature*. Rockefeller University Press, New York, 1978, p. 97.

38. Weissmann, C., interview in Hargittai, I., *Candid Science II: Conversations with Famous Biomedical Scientists*. Imperial College Press, London, 2002, pp. 466–97.

39. Gajdusek, D. C., interview in Hargittai, I., *Candid Science II: Conversations with Famous Biomedical Scientists*. Imperial College Press, London, 2002, pp. 442–65.

40. This was a nice example of Karl Popper's ideas of scientific method.

41. Caroe, G. M., *William Henry Bragg, 1862–1942: Man and Scientist*. Cambridge University Press, Cambridge and London, 1978.

42. Letter of J. D. Bernal, 27 January 1948, Nobel Archives.

43. Letter of H. C. Urey, 4 April 1947, Nobel Archives.

44. Letters of H. C. Urey, 5 April and 25 April 1947, Nobel Archives.

45. Bernstein, J., *Hitler's Uranium Club: The Secret Recordings at Farm Hall*. American Institute of Physics, Woodburg, New York, 1996.

46. Kroto, H. W.; Heath, J. R.; O'Brien, S. C.; Curl, R. F.; Smalley, R. E., *Nature*, 1985, **318**, 162–3.

47. Later it became clear that not all the graduate students had got their names onto the paper, not by design but by oversight.

48. Curl, R. F., interview in Hargittai, I., *Candid Science: Conversations with Famous Chemists*. Imperial College Press, London, pp. 374–84.

49. Henderson, R., private communication, 2001.

50. Deisenhofer, J.; Epp, O.; Miki, K.; Huber, R.; Michel, H., *J. Mol. Biol.*, 1984, 180, 385–98; *Nature*, 1985, 318, 618–24.

51. Michel, H., interview in Hargittai, I., *Candid Science III: More Conversations with Famous Chemists*. Imperial College Press, London, 2003, pp. 332–41.

52. Deisenhofer, J., interview in Hargittai, I., *Candid Science III: More Conversations with Famous Chemists*. Imperial College Press, London, 2003, pp. 342–53.

53. Huber, R., in *Nobel Lectures: Chemistry 1981–1990*. Editor-in-charge T. Frängsmyr, editor B. G. Malström. World Scientific, Singapore, 1992, p. 567.

54. Huber, R., interview in Hargittai, I., *Candid Science III: More Conversations with Famous Chemists*. Imperial College Press, London, 2003, pp. 354–67.

55. Medawar, P., *The Threat and the Glory: Reflections on Science and Scientists*. Oxford University Press, 1991, p. 242.

56. Zydowsky, T. M., *Chem. Intell.*, 2000, 6(1), 29–34.

57. Lindqvist, I., in *Nobel Lectures: Chemistry 1971–1980*. Editor-in-charge T. Frängsmyr, editor S. Forsén. World Scientific, Singapore, 1993, p. 99.

58. Wilkinson, G., in *Nobel Lectures: Chemistry 1971–1980*. Editor-in-charge T. Frängsmyr, editor S. Forsén. World Scientific, Singapore, 1993, pp. 137–45.

59. Robbins, F. C., conversation with M. Hargittai, Cleveland, Ohio, October 2000, in Hargittai, I., *Candid Science II: Conversations with Famous Biomedical Scientists*. Imperial College Press, London, 2002, pp. 498–517.

60. Enders, J. F.; Weller, T. H. E.; Robbins, F. C. *Science*, 1949, 109, 85–7.

61. Robbins, F. C., *J. Pediatrics*, 1994, 124, 155–7, p. 156.

62. I appreciate Kerstin Fredga's kindness in letting me quote her on this story It was on the day of the announcements of the physics and chemistry prizes in 1996. The announcements were followed by a lecture, which I gave. My topic happened to be symmetry, which was very appropriate as both the physics and chemistry prizes that year were related to the symmetry concept. A dinner at the Academy concluded the day's events. It was obvious that the members of the

Academy were relaxed after another strenuous week of heightened Nobel activities. The president, astrophysicist Kerstin Fredga, gave a brief speech at this dinner party. Her story has been in circulation in other versions, and it probably emulated a political anecdote about Bill Clinton.

63. Lubkin, G. B., *Physics Today*, April 1992, p. 24.

64. Olah, G. A., interview in Hargittai, I., *Candid Science: Conversations with Famous Chemists*. Imperial College Press, London, 2000, pp. 270–83.

65. Peter Mitchell gives the sources of this statement: Mitchell, P., *Nobel Lectures: Physiology or Medicine 1971–1980*. World Scientific, Singapore, 1998, pp. 295, 329.

66. Crick, F., *What Mad Pursuit: A Personal View of Scientific Discovery*. Basic Books, New York, 1988, p. 70.

67. Berg, P., interview in Hargittai, I., *Candid Science II: Conversations with Famous Biomedical Scientists*. Imperial College Press, London, 2002, pp. 154–81.

68. Barton, D. H. R., interview in Hargittai, I., *Candid Science: Conversations with Famous Chemists*. Imperial College Press, London, 2000, pp. 148–57.

69. Cram, D. J., interview in Hargittai, I., *Candid Science III: More Conversations with Famous Chemists*. Imperial College Press, London, 2003, pp. 178–97.

70. Brown, H. C., interview in Hargittai, I., *Candid Science: Conversations with Famous Chemists*. Imperial College Press, London, 2000, pp. 250–69.

71. Letter of M. S. Livingston to C. J. Davisson, 3 January 1939, Nobel Archives.

72. Wideroe, R., *Arch. f. Elektrotechnik*, 1928, **21**, 387.

73. Kornberg, A., Nobel essay on the web site of the Nobel e-Museum, http://www.nobel.se/essays/research/index.html

74. Feldman, B., *The Nobel Prize: A History of Genius, Controversy, and Prestige*. Arcade Publishing, New York, 2000, pp. 286–9.

75. Elion, G. B., in Hargittai, I., *Candid Science: Conversations with Famous Chemists*. Imperial College Press, London, 2000, pp. 54–71.

76. Saxon, W., *The New York Times*, 22 December 1998, p. C19.

77. Jackson, J. D.; Panofsky, W. K. H., *Biographical Memoirs of the National Academy of Sciences of the U. S. A.* (on Edwin McMillan), Volume 69. National Academy Press, Washington, DC, 1996, pp. 215–40.

78. Glashow, S. with Ben Bova, *Interactions: A Journey through the Mind of a Particle Physicist and the Matter of This World*. Warner Books, New York, 1988, pp. 265–71.

79. Perutz, M. F., *I Wish I'd Made You Angry Earlier: Essays on Science and Scientists*. Oxford University Press, 1998, p. 298.

80. Ferry, G., *Dorothy Hodgkin: A Life*. Granta Books, London, 1998, pp. 286–9.

81. This is according to Ferry, G., *Dorothy Hodgkin: A Life*, Granta Books, London, 1998, pp. 289, 414, on the basis of the Papers of W. L. Bragg in the Royal Institution.

82. See, for example, *Forum: Journal of the Association for Women in Science and Engineering*, No. 8, 1999/2000, pp. 2–3. The percentages for science academies vary from 14.6% in the Turkish Academy of Sciences to 0.4% in the Royal Netherlands Academy of Arts and Sciences. The United States National Academy of Sciences, the Royal Society (London), the Royal Swedish Academy of Sciences, and the Japan Academy were at 6.2, 3.6, 5.5, and 0.8%, respectively, according to the mostly 1999 data.

83. Straus, E., *Rosalyn Yalow, Nobel Laureate: Her Life and Work in Medicine, A Biographical Memoir*. Plenum, New York and London, 1998, pp. 240–1.

84. Boyer, P. D., interview in Hargittai, I., *Candid Science III: More Conversations with Famous Chemists*. Imperial College Press, London, 2003, pp. 268–79.

85. Ruska, E., in *Nobel Lectures: Physics 1981–1990*. World Scientific, Singapore, 1993, p. 355.

86. Kapitsa, P. L., in *Nobel Lectures: Physics 1971–1980*. World Scientific, Singapore, 1992, p. 424.

87. Letter of nomination by P. A. M. Dirac, 16 January 1946, Nobel Archives.

88. Letter of nomination by C. G. Darwin, 5 January 1940, Nobel Archives.

89. Lederberg, J., interview in Hargittai, I., *Candid Science II: Conversations with Famous Biomedical Scientists*. Imperial College Press, London, 2002, pp. 32–49.

90. Avery, O. T.; MacLeod, C.; McCarty, M., *J. Exp. Med.*, 1944, **79**, 137–58.

91. Lederberg, J., in *Nobel Lectures: Physiology or Medicine 1942–1962*. World Scientific, Singapore, 1999, p. 620.

92. Ochoa, S., in *Nobel Lectures: Physiology or Medicine 1942–1962*. World Scientific, Singapore, 1999, p. 661.

93. Kornberg, A., in *Nobel Lectures: Physiology or Medicine 1942–1962*. World Scientific, Singapore, 1999, pp. 666, 667.

94. Hauptman, H. A., interview in Hargittai, I., *Candid Science III: More Conversations with Famous Chemists*. Imperial College Press, London, 2003, pp. 292–317.

95. Smalley, R. E., interview in Hargittai, I., *Candid Science: Conversations with Famous Chemists*. Imperial College Press, London, 2000, pp. 362–73.

4 Discoveries

1. Klug, A., interview in Hargittai, I., *Candid Science II: Conversations with Famous Biomedical Scientists*. Imperial College Press, London, 2002, pp. 306–29.

2. Dubos, R., *Louis Pasteur: Free Lance to Science*. Da Capo Press, New York, 1986, p. 369.

3. Root-Bernstein, R., *Discovering*. Harvard University Press, Cambridge, Massachusetts, and London, 1991, pp. xii–xiii.

4. Feynman, R. P., in *Nobel Lectures: Physics 1963–1970*. World Scientific, Singapore, 1998, p.155.

5. Quoted by David Pyke, Introduction to Medawar, P., *The Threat and the Glory: Reflections on Science and Scientists*. Oxford University Press, 1991, p. xvi.

6. Jacob, F., interview in Hargittai, I., *Candid Science II: Conversations with Famous Biomedical Scientists*. Imperial College Press, London, 2002, pp. 84–97.

7. Crick, F., *What Mad Pursuit: A Personal View of Scientific Discovery*. Basic Books, New York, 1988, p. 76.

8. Segrè, E. (on the basis of Clarence Larson's interview), in Hargittai, I.; Hargittai, M., *Chem. Intell.*, 2000, 6(4), 54–6.

9. Emsley, J., quoting an interview in *C&EN, Chemistry in Britain*, November 1997, pp. 43–5.

10. Lederberg, J., interview in Hargittai, I., *Candid Science II: Conversations with Famous Biomedical Scientists*. Imperial College Press, London, 2002, pp. 32–49.

11. Lederberg decided to limit his involvement to providing the tremendous inventory of all his documents on his web site and letting future analysts deal with the problem (see previous note).

12. Apparently the well–established Severo Ochoa (M59) was also on the right track but did not have the serendipity that helped Nirenberg and Matthaei so much.

13. Edelman, G. M., interview in Hargittai, I., *Candid Science II: Conversations with Famous Biomedical Scientists*. Imperial College Press, London, 2002, pp. 196–219.

14. Ramdas, L. A., *J. Phys. Educ.*, 1973, 1(3), 2–18, as quoted in Mallik, D. C. V., *Notes Rec. R. Soc. London*, 2000, 54(1), 67–83, p. 72.

15. Hargittai, M. Conversation with Frederick Robbins, Cleveland, Ohio, October 2000, in Hargittai, I., *Candid Science II: Conversations with Famous Biomedical Scientists*. Imperial College Press, London, 2002, pp. 498–517.

16. Berry, R. S., interview in Hargittai, I., *Candid Science: Conversations with Famous Chemists*. Imperial College Press, London, 2000, pp. 422–35.

17. Berry, R. S., *J. Chem. Phys.*, 1960, 32, 933–8.

18. Hargittai, I., *Chem. Intell.*, 1997, 3(4), 25–49.

19. Shechtman, D.; Blech, I.; Gratias, D.; Cahn, J. W., *Phys. Rev. Lett.*, 1984, 53, 1951–3.

20. Eaton, P. E., interview in Hargittai, I., *Candid Science: Conversations with Famous Chemists*. Imperial College Press, London, 2000, pp. 416–21.

21. Kroto, H. W., interview in Hargittai, I., *Candid Science: Conversations with Famous Chemists*. Imperial College Press, London, 2000, pp. 332–57.

22. Osawa, E., interview in Hargittai, I., *Candid Science: Conversations with Famous Chemists*. Imperial College Press, London, 2000, pp. 308–21.

23. Kroto, H. W.; Heath, J. R.; O'Brien, S. C.; Curl, R. F.; Smalley, R. E., *Nature*, 1985, 318, 162–3.

24. Krätschmer, W.; Lamb, L. D.; Fostiropoulos, K.; Huffman, D. R., *Nature*, 1990, 347, 354–8.

25. Mullis, K. B., interview in Hargittai, I., *Candid Science II: Conversations with Famous Biomedical Scientists*. Imperial College Press, London, 2002, pp. 182–95.

26. Rowland, F. S., interview in Hargittai, I., *Candid Science: Conversations with Famous Chemists*. Imperial College Press, London, 2000, pp. 448–65.

27. Ernster, L., interview in Hargittai, I., *Candid Science II: Conversations with Famous Biomedical Scientists*. Imperial College Press, London, 2001, pp. 376–95.

28. Vladimir Prelog's statement of 1995 in Eschenmoser, A., interview in Hargittai, I., *Candid Science III: More Conversations with Famous Chemists*. Imperial College Press, London, 2003, pp. 96–107.

29. Berg, P., interview in Hargittai, I., *Candid Science II: Conversations with Famous Biomedical Scientists*. Imperial College Press, London, 2002, pp. 154–81.

30. Watson, J. D.; Tooze, J., *The DNA Story: A Documentary History of Gene Cloning*. W. H. Freeman, San Francisco, 1981.

31. Perutz, M., *Acta Cryst.*, 1970, A 26, 183–5, p. 184.

32. Bernal, J. D., *Biographical Memoirs of Fellows of the Royal Society* (on William Astbury), November 1963, 9, 1–35, p. 27.

33. Dagani, R., *C&EN*, 23 May 1994, 39–42, p. 40.

34. Late in his career, though, Szent-Györgyi made some unsubstantiated claims regarding the cure of cancer, but this was not characteristic of much of his scientific activities for decades.

35. Conversations with Ferenc Guba, a former associate of Szent-Györgyi in Szeged, Hungary, and with Andrew Szent-Györgyi, nephew and former co-worker with his uncle in Woods Hole. These conversations took place in Budapest in 1999, unpublished records.

36. Medawar, P., *The Art of the Soluble*. Methuen, London, 1967, p. 7.

37. Calvin, M. (on the basis of Clarence Larson's interview), in Hargittai, I.; Hargittai, M., *Chem. Intell.*, 2000, 6(1), 52–5.

38. Rohlfing, E. A.; Cox, D. M.; Kaldor, A., *J. Chem. Phys.*, 1984, 81, 3322–30.

39. Whetten, R. L., interview in Hargittai, I., *Candid Science: Conversations with Famous Chemists*. Imperial College Press, London, 2000, pp. 405–15.

40. Barton, D. H. R., interview in Hargittai, I., *Candid Science: Conversations with Famous Chemists*. Imperial College Press, London, 2000, pp. 148–57.

41. Szent-Györgyi, A., *Bioenergetics*. Academic Press, New York, 1957.

42. Boyer, P. D., interview in Hargittai, I., *Candid Science III: More Conversations with Famous Chemists*. Imperial College Press, London, 2003, pp. 268–79.

43. Fischer, E. P.; Lipson, C., *Thinking about Science: Max Delbruck and the Origins of Molecular Biology*. W. W. Norton, New York, 1988.

44. Gillespie, R. J., interview in Hargittai, I., *Candid Science III: More Conversations with Famous Chemists*. Imperial College Press, London, 2003, pp. 48–57.

45. Pauling, L., *Chem. Intell.* 1996, **2**(1), 32–8. This was a posthumous publication by Linus Pauling, communicated by his long-time secretary Dorothy Munro.

46. Crick, F., *What Mad Pursuit: A Personal View of Scientific Discovery*. Basic Books, New York, 1988, p. 59.

47. Lee, T. D., *Nature*, 1997, **386**, 334.

48. Glashow, S. with Ben Bova, *Interactions: A Journey Through the Mind of a Particle Physicist and the Matter of this World*. Warner Books, New York, 1988, p. 128.

49. Pople, J. A., interview in Hargittai, I., *Candid Science: Conversations with Famous Chemists*. Imperial College Press, London, 2000, pp. 178–89.

50. Crick, F., *What Mad Pursuit: A Personal View of Scientific Discovery*. Basic Books, New York, 1988, p. 78.

51. Olah, G. A., interview in Hargittai, I., *Candid Science: Conversations with Famous Chemists*. Imperial College Press, London, 2000, pp. 270–83.

52. Sanger, F., interview in Hargittai, I., *Candid Science II: Conversations with Famous Biomedical Scientists*. Imperial College Press, London, 2002, pp. 72–83.

53. Kornberg, A., *For the Love of Enzymes*. Harvard University Press, Cambridge, Massachusetts, and London, 1989, p. 139.

54. Khorana, H. G., *Chemical Biology: Selected Papers of H. Gobind Khorana (with introductions)*. World Scientific, Singapore, 2000.

55. Cram, D. J., interview in Hargittai, I., *Candid Science III: More Conversations with Famous Chemists*. Imperial College Press, London, 2003, pp. 178–97.

56. Hey, A. J. G., *Physics Today*, September 1996, 44–9, p. 49.

57. Ramaseshan, S., *Current Science*, 1997, **72**, 423–6.

58. Dunitz, J. D., *Current Science*, 1997, **72**, 447–50.

59. Cornforth, J. W., in *Organic Reactivity: Physical and Biological Aspects*. Edited by B. T. Golding, R. J. Griffin, and H. Maskill. Royal Society of Chemistry, Cambridge, 1995, pp. 25–37.

60. Perutz, M., *Current Science* 1997, **72**, 450–3.

61. Cornforth, J. W., private communication, 1998.

62. Nirenberg, M. W., profile in Hargittai, I., *Candid Science II: Conversations with Famous Biomedical Scientists*. Imperial College Press, London, 2002, pp. 130–41.

63. Altman, S., interview in Hargittai, I., *Candid Science II: Conversations with Famous Biomedical Scientists*. Imperial College Press, London, 2002, pp. 338–49.

64. Dubos, R., *Louis Pasteur: Free Lance to Science*. Da Capo Press, New York, 1986, p. 95.

65. Pasteur, L., *Researches on the Molecular Asymmetry of Natural Organic Products*. Alembic Club Reprints, No. 14. W. F. Clay, Edinburgh, 1897, p. 21.

66. Deisenhofer, J., interview in Hargittai, I., *Candid Science III: More Conversations with Famous Chemists*. Imperial College Press, London, 2003, pp. 342–53.

67. Szent-Györgyi, A., *Ann. Rev. Biochem.*, 1963, **32**, 1–14, p. 9.

68. Glashow, S. with Ben Bova, *Interactions: A Journey Through the Mind of a Particle Physicist and the Matter of this World*. Warner Books, New York, 1988, p. 60.

69. Lubkin, G. B., *Physics Today*, April 1992, p. 23.

70. Gajdusek, D. C., interview in Hargittai, I., *Candid Science II: Conversations with Famous Biomedical Scientists*. Imperial College Press, London, 2002, pp. 442–65.

5 Overcoming adversity

1. Lederman, L., interview in Hargittai, I.; Hargittai, M., *Chem. Intell.*, 1998, **4**(4), 20–9.

2. Cram, D. J., interview in Hargittai, I., *Candid Science III: More Conversations with Famous Chemists*. Imperial College Press, London, 2003, pp. 178–97.

3. Gilman, A. G., interview in Hargittai, I., *Candid Science II: Conversations with Famous Biomedical Scientists*. Imperial College Press, London, 2002, pp. 238–51.

4. Goodman, L. S.; Gilman, A., *The Pharmacological Basis of Therapeutics*. Macmillan, New York, 1941 (the year when Alfred Gilman Jr. was born).

5. Gilman, A. G., in *Nobel Lectures: Physiology or Medicine 1991–1995*. Edited by N. Ringertz. World Scientific, Singapore, 1997, p. 179.

6. Gajdusek, D. C., interview in Hargittai, I., *Candid Science II: Conversations with Famous Biomedical Scientists*. Imperial College Press, London, 2002, pp. 442–65.

7. Merrifield, B., interview in Hargittai, I., *Candid Science III: More Conversations with Famous Chemists*. Imperial College Press, London, 2003, pp. 206–19.

8. Nathans, D., interview in Hargittai, I., *Candid Science II: Conversations with Famous Biomedical Scientists*. Imperial College Press, London, 2002, pp. 142–53.

9. Brown, H. C., interview in Hargittai, I., *Candid Science: Conversations with Famous Chemists*. Imperial College Press, London, 2000, pp. 250–69.

10. The same Gordon who then initiated the popular Gordon Research Conferences in New England.

11. Elion, G. B., interview in Hargittai, I., *Candid Science: Conversations with Famous Chemists*. Imperial College Press, London, 2000, pp. 54–71.

12. Bergström, K. S. D., profile in Hargittai, I., *Candid Science II: Conversations with Famous Biomedical Scientists*. Imperial College Press, London, 2002, pp. 542–7.

13. Barton, D. H. R., interview in Hargittai, I., *Candid Science: Conversations with Famous Chemists*. Imperial College Press, London, 2000, pp. 148–57.

14. Porter, G., interview in Hargittai, I., *Candid Science: Conversations with Famous Chemists*. Imperial College Press, London, 2000, pp. 476–87.

15. Black, J. W., profile in Hargittai, I., *Candid Science II: Conversations with Famous Biomedical Scientists*. Imperial College Press, London, 2002, pp. 524–41.

16. Walker, J. E., interview in Hargittai, I., *Candid Science III: More Conversations with Famous Chemists*. Imperial College Press, London, 2003, pp. 280–91.

17. Semenov, N. N., interview in Hargittai, I., *Candid Science: Conversations with Famous Chemists*. Imperial College Press, London, 2000, pp. 466–75.

18. Nirenberg, M. W., profile in Hargittai, I., *Candid Science II: Conversations with Famous Biomedical Scientists*. Imperial College Press, London, 2002, pp. 130–41.

19. Cornforth, J. W., interview in Hargittai, I., *Candid Science: Conversations with Famous Chemists*. Imperial College Press, London, pp. 122–37.

20. Taube, H., interview in Hargittai, I., *Candid Science III: More Conversations with Famous Chemists*. Imperial College Press, London, 2003, pp. 400–13.

21. Prelog, V., interview in *Candid Science: Conversations with Famous Chemists*. Imperial College Press, London, 2000, pp. 138–47.

22. Hoffmann, R., interview in Hargittai, I., *Candid Science: Conversations with Famous Chemists.* Imperial College Press, London, 2000, pp. 190–209.

23. Emsley, J., *Chemistry in Britain*, November 1997, pp. 43–5.

24. Blobel, G., interview in Hargittai, I., *Candid Science II: Conversations with Famous Biomedical Scientists.* Imperial College Press, London, 2002, pp. 252–65.

25. Brown, A., *The Neutron and the Bomb: A Biography of Sir James Chadwick.* Oxford University Press, 1997.

26. Jacob, F., interview in Hargittai, I., *Candid Science II: Conversations with Famous Biomedical Scientists.* Imperial College Press, London, 2002, pp. 84–97.

27. Hargittai, M., conversation with Rita Levi-Montalcini in Rome in June 2000; in Hargittai, I., *Candid Science II: Conversations with Famous Biomedical Scientists.* Imperial College Press, London, 2002, pp. 364–75.

28. Levi-Montalcini, R., *In Praise of Imperfection: My Life and Work.* Basic Books, New York, 1988, pp. 94–5.

29. Krebs, H. A., *Nobel Lectures: Physiology or Medicine 1942–1962.* World Scientific, Singapore, 1999, p. 411.

30. Herzberg, G., in *Nobel Lectures: Chemistry 1971–1980.* Editor-in-charge T. Frängsmyr, editor S. Forsén. World Scientific, Singapore, 1993, p. 7.

31. Hauptman, H. A., interview in Hargittai, I., *Candid Science III: More Conversations with Famous Chemists.* Imperial College Press, London, 2003, pp. 292–317.

32. Furchgott, R. F., interview in Hargittai, I., *Candid Science II: Conversations with Famous Biomedical Scientists.* Imperial College Press, London, 2002, pp. 578–93.

33. Kornberg, A., interview in Hargittai, I., *Candid Science II: Conversations with Famous Biomedical Scientists.* Imperial College Press, London, 2002, pp. 50–71.

34. Karle, J., conversation in Washington, DC, in 2000, unpublished records.

35. Yalow, R., interview in Hargittai, I., *Candid Science II: Conversations with Famous Biomedical Scientists.* Imperial College Press, London, 2002, pp. 518–23.

36. Berg, P., interview in Hargittai, I., *Candid Science II: Conversations with Famous Biomedical Scientists.* Imperial College Press, London, 2002, pp. 154–81.

37. Lederman, L., interview in Hargittai, I.; Hargittai, M., *Chem. Intell.*, 1998, 4(4), 20–9.

38. Varmus, H. E., in *Nobel Lectures: Physiology or Medicine 1981–1990*. Editor-in-charge T. Frängsmyr, editor J. Lindsten. World Scientific, Singapore, 1993. p. 501.

39. Cohen, S., in *Nobel Lectures: Physiology or Medicine 1981–1990*. Editor-in-charge T. Frängsmyr, editor J. Lindsten. World Scientific, Singapore, 1993, p. 331.

40. Milstein, C., interview in Hargittai, I., *Candid Science II: Conversations with Famous Biomedical Scientists*. Imperial College Press, London, 2002, pp. 220–37.

41. Olah, G. A., interview in Hargittai, I., *Candid Science: Conversations with Famous Chemists*. Imperial College Press, London, 2000, pp. 270–83.

42. Tsui, D. C., conversation in Princeton, 1999, unpublished records.

43. Zewail, A. H., interview in Hargittai, I., *Candid Science: Conversations with Famous Chemists*. Imperial College Press, London, 2000, pp. 488–507.

44. Boag, J. W.; Rubinin, P. E.; Shoenberg, D., *Kapitza in Cambridge and Moscow: Life and Letters of a Russian Physicist*. North-Holland, Amsterdam, 1990.

45. Later they changed the rules of the Royal Society so that a foreign national could not become a fellow, only a foreign member. Now it is again possible.

46. Shoenberg, D., interview in Hargittai, I., *Chem. Intell.*, 2000, 6(4), 14–19.

47. Gorelik, G., *Physics Today*, May 1995, 11–15, p. 86.

6 *What turned you to science?*

1. I recently visited the collection of old chemistry sets at the Department of Chemistry, University of Cincinnati and was impressed by their variety and attractive appearance.

2. De Kruif, P., *Microbe Hunters*. Harcourt, Brace & World, New York, 1996.

3. De Kruif published a personal account of his life in 1962, *The Sweeping Wind: A Memoir* (Harcourt, Brace & World, New York, 1962), which contains interesting information about the circumstances of writing *Microbe Hunters*.

4. Lewis, S., *Arrowsmith*. Buccaneer Books, Cutchogue, New York, 1996. De Kruif contributed the science-related aspects to *Arrowsmith*. He writes about this in detail in *The Sweeping Wind*.

5. Berg, P., interview in Hargittai, I., *Candid Science II: Conversations with Famous Biomedical Scientists*. Imperial College Press, London, 2002, pp. 154–81.

6. Hauptman, H. A., interview in Hargittai, I., *Candid Science III: More Conversations with Famous Chemists*. Imperial College Press, London, 2003, pp. 292–317.

7. Semenov, N. N., interview in Hargittai, I., *Candid Science: Conversations with Famous Chemists*. Imperial College Press, London, 2000, pp. 466–75.

8. Lipscomb, W. N., interview in Hargittai, I., *Candid Science III: More Conversations with Famous Chemists*. Imperial College Press, London, 2003, pp. 18–27.

9. Boyer, P. D., interview in Hargittai, I., *Candid Science III: More Conversations with Famous Chemists*. Imperial College Press, London, 2003, pp. 268–79.

10. Curl, R. F., interview in Hargittai, I., *Candid Science: Conversations with Famous Chemists*. Imperial College Press, London, 2000, pp. 374–84.

11. Olah, G. A., interview in Hargittai, I., *Candid Science: Conversations with Famous Chemists*. Imperial College Press, London, 2000, pp. 270–83.

12. Lederman, L., interview in Hargittai, I.; Hargittai, M., *Chem. Intell.*, 1998, 4(4), 20–9.

13. Vane, J. R., interview in Hargittai, I., *Candid Science II: Conversations with Famous Biomedical Scientists*. Imperial College Press, London, 2002, pp. 548–63.

14. Stevenson, R., *Chemistry in Britain*, June 1999, pp. 20–2.

15. Porter, G., interview in Hargittai, I., *Candid Science: Conversations with Famous Chemists*. Imperial College Press, London, 2000, pp. 476–87.

16. Walker, J. E., interview in Hargittai, I., *Candid Science III: More Conversations with Famous Chemists*. Imperial College Press, London, 2003, pp. 280–91.

17. Watson, J. D., interview in Hargittai, I., *Candid Science II: Conversations with Famous Biomedical Scientists*. Imperial College Press, London, 2002, pp. 2–15.

18. Lehn, J.-M., interview in Hargittai, I., *Candid Science III: More Conversations with Famous Chemists.* Imperial College Press, London, 2003, pp. 198–206.

19. Although many think that it was one particular high school where all the famous Hungarian scientists went, there were several very strong high schools. Between them, Dennis Gábor, John Harsányi, George Hevesy, Theodore Kármán, John Neumann, George Oláh, Michael Polányi, Leo Szilárd, Edward Teller, and Eugene Wigner went to a total of five different high schools in Budapest. Around the turn of the twentieth century, religious affiliation in choosing a high school was much less important than it became later. Thus, for example, in 1915, 28% of the students of the Lutheran high school were actually Lutheran and 53% were Jewish. At that time the high school teacher had a respectable social status, was well paid, and was usually male. Only a relatively small, select fraction of the general population attended high school.

20. My encounters with Wigner span three decades. I first started corresponding with him when I published a note in a Budapest literary weekly in 1964 concerning his ideas about the limits of science. He sent me a nice letter in response. We met in person only once, in 1969. He spent a week in the Department of Physics at the University of Texas and I was spending a year in that department as a research associate. Wigner dropped by my office every morning for a one-hour chat before his official program started.

21. Students attended these high schools for eight years, from ages 11 through 18, which would correspond to the 5th to 12th grades.

22. This teacher also recognized Neumann's talent.

23. Wigner, E. P., in *Symmetries and Reflections: Scientific Essays.* Indiana University Press, Bloomington and London, 1963, pp. 262–3.

24. Gabor, D., in *Nobel Lectures: Physics 1971–1980.* World Scientific, Singapore, 1992, pp. 7–9.

25. Seaborg, G. T., interview in Hargittai, I., *Candid Science III: More Conversations with Famous Chemists.* Imperial College Press, London, 2003, pp. 2–17.

26. Cornforth, C. W., interview in Hargittai, I., *Candid Science: Conversations with Famous Chemists.* Imperial College Press, London, 2000, pp. 122–37.

27. Black, J. W., profile in Hargittai, I., *Candid Science II: Conversations with Famous Biomedical Scientists.* Imperial College Press, London, 2002, pp. 524–41.

28. Brown, H. C., interview in Hargittai, I., *Candid Science: Conversations with Famous Chemists*. Imperial College Press, London, 2000, pp. 250–69.

29. Bergström, K. S. D., profile in Hargittai, I., *Candid Science II: Conversations with Famous Biomedical Scientists*. Imperial College Press, London, 2002, pp. 542–7.

30. Fukui, K., interview in Hargittai, I., *Candid Science: Conversations with Famous Chemists*. Imperial College Press, London, 2000, pp. 210–21.

31. Klug, A., interview in Hargittai, I., *Candid Science II: Conversations with Famous Biomedical Scientists*. Imperial College Press, London, 2002, pp. 306–29.

32. Lederberg, J., interview in Hargittai, I., *Candid Science II: Conversations with Famous Biomedical Scientists*. Imperial College Press, London, 2002, pp. 32–49.

33. Hoffmann, R., interview in Hargittai, I., *Candid Science: Conversations with Famous Chemists*. Imperial College Press, London, 2000, pp. 190–209.

34. Lewis, E. B., interview in Hargittai, I. *Candid Science II: Conversations with Famous Biomedical Scientists*. Imperial College Press, London, 2002, pp. 350–63.

35. Gajdusek, D. C., interview in Hargittai, I., *Candid Science II: Conversations with Famous Biomedical Scientists*. Imperial College Press, London, 2002, pp. 442–65.

36. Smalley, R. E., interview in Hargittai, I., *Candid Science: Conversations with Famous Chemists*. Imperial College Press, London, 2000, pp. 362–73.

37. Mullis, K. B., interview in Hargittai, I., *Candid Science II: Conversations with Famous Biomedical Scientists*. Imperial College Press, London, 2002, pp. 182–95.

38. Elion, G. B., interview in Hargittai, I., *Candid Science: Conversations with Famous Chemists*. Imperial College Press, London, 2000, pp. 54–71.

39. Segrè, E. (based on Clarence Larson's interview), Hargittai, I.; Hargittai, M., *Chem. Intell.* 2000, 6(4), 54–6.

40. Chamberlain, O., conversation in Berkeley, California, 1999, unpublished records.

41. Prigogine, I., interview in Hargittai, I., *Candid Science III: More Conversations with Famous Chemists*. Imperial College Press, London, 2003, pp. 422–31.

42. Sanger, F., interview in Hargittai, I., *Candid Science II: Conversations with Famous Biomedical Scientists*. Imperial College Press, London, 2002, pp. 72–83.

43. Deisenhofer, J., interview in Hargittai, I., *Candid Science III: More Conversations with Famous Chemists*. Imperial College Press, London, 2003, pp. 342–53.

44. Nirenberg, M. W., profile in Hargittai, I., *Candid Science II: Conversations with Famous Biomedical Scientists*. Imperial College Press, London, 2002, pp. 130–41.

45. Zewail, A. H., interview in Hargittai, I., *Candid Science: Conversations with Famous Chemists*. Imperial College Press, 2000, pp. 488–507.

7 *Venue*

1. Mackay, A. L., *Chem. Intell.*, 1995, 1(1), 12–18, p. 14.

2. Watson, J. D., *A Passion for DNA: Genes, Genomes, and Society*. Oxford University Press, 2000, p. 124.

3. Gajdusek, D. C., interview in Hargittai, I., *Candid Science II: Conversations with Famous Biomedical Scientists*. Imperial College Press, London, 2002, pp. 442–65.

4. Watson, J. D., *A Passion for DNA: Genes, Genomes, and Society*. Oxford University Press, 2000, pp. 4–5.

5. Watson, J. D., in *Phage and the Origin of Molecular Biology*. Edited by J. Cairns, G. Stent, and J. D. Watson. Cold Spring Harbor Laboratory of Quantitative Biology, 1966, p. 239.

6. Black, J. W., profile in Hargittai, I., *Candid Science II: Conversations with Famous Biomedical Scientists*. Imperial College Press, London, 2002, 524–41.

7. According to a note in the MRC LMB Archives.

8. Henderson, R., interview in Hargittai, I., *Candid Science II: Conversations with Famous Biomedical Scientists*. Imperial College Press, London, 2002, 296–305.

9. Watson, J. D., *A Passion for DNA: Genes, Genomes, and Society*. Oxford University Press, 2000, p. 22.

10. Bernal, J. D., *Biographical Memoirs of Fellows of the Royal Society* (on William Astbury), November 1963, 9, 1–35, pp. 6–7.

11. Altman, S., interview in Hargittai, I., *Candid Science II: Conversations with Famous Biomedical Scientists*. Imperial College Press, London, 2002, pp. 338–49.

12. Müller-Hill, B., interview in Hargittai, I., *Candid Science II: Conversations with Famous Biomedical Scientists*. Imperial College Press, London, 2002, 114–29.

13. Oliphant, M., *Rutherford: Recollections of the Cambridge Days*. Elsevier, Amsterdam, 1972, pp. 42–5.

14. Fuller, M., conversation in Cambridge, 2000, unpublished records.

15. This seems to contradict Rutherford's dictum, 'Never attempt a difficult problem' (cf. Chapter 4, p. 89). In Sanger's words difficult means challenging, and of course the difficulty of a problem is relative. Sanger attacked enormous problems, which were, however, solvable.

16. Walker, J. E., interview in Hargittai, I., *Candid Science III: More Conversations with Famous Chemists*. Imperial College Press, London, 2003, pp. 280–91.

17. Perutz, M., *Acta Cryst.*, 1970, A **26**, 183–5, p. 185.

18. Ryle, M., in *Nobel Lectures: Physics 1971–1980*. World Scientific, Singapore, 1992, p. 185.

19. Cohen, M., *Chem. Intell.*, 1997, 3(2), 33–40.

20. Brown, A., *The Neutron and the Bomb: A Biography of Sir James Chadwick*. Oxford University Press, 1997.

21. Oliphant, M., interview in Hargittai, I.; Hargittai, M., *Chem. Intell.*, 2000, **6**(3), 50–4.

22. Chadwick, J., Foreword to Oliphant, M., *Rutherford: Recollections of the Cambridge Days*. Elsevier, Amsterdam, 1972, p. X.

23. Heilbron, J. L.; Seidel, R. W., *Lawrence and His Laboratory: A History of the Lawrence Berkeley Laboratory*. Vol. 1. University of California Press, Berkeley, 1989, p. 489.

24. Heilbron, J. L.; Seidel, R. W., *Lawrence and His Laboratory: A History of the Lawrence Berkeley Laboratory*. Vol. 1. University of California Press, Berkeley, 1989, p. 226.

25. Boag, J. W.; Rubinin, P. E.; Shoenberg, D., *Kapitza in Cambridge and Moscow: Life and Letters of a Russian Physicist*. North-Holland, Amsterdam, 1990, p. 71.

26. Perutz, M., interview in Hargittai, I., *Candid Science II: Conversations with Famous Biomedical Scientists*. Imperial College Press, London, 2002, pp. 280–95.

27. Furchgott, R. F., interview in Hargittai, I., *Candid Science II: Conversations with Famous Biomedical Scientists*. Imperial College Press, London, 2002, pp. 578–93.

28. Jacob, F., interview in Hargittai, I., *Candid Science II: Conversations with Famous Biomedical Scientists.* Imperial College Press, London, 2002, pp. 84–97.

29. Levi-Montalcini, R., *In Praise of Imperfection: My Life and Work.* Basic Books, New York, 1988, p. 195.

30. Merrifield, B., interview in Hargittai, I., *Candid Science III: More Conversations with Famous Chemists.* Imperial College Press, London, 2003, pp. 206–20.

31. Maddox, J., interview in Hargittai, I., *Chem. Intell.*, 1999, 5(2), 53–5.

32. Taube, H., interview in Hargittai, I., *Candid Science III: More Conversations with Famous Chemists.* Imperial College Press, London, 2003, pp. 400–13.

33. Cram, D. J., interview in Hargittai, I., *Candid Science III: More Conversations with Famous Chemists.* Imperial College Press, London, 2003, pp. 178–97.

34. Corey, E. J., in *Nobel Lectures: Chemistry 1981–1990.* Editor-in-charge T. Frängsmyr, editor B. G. Malmström. World Scientific, Singapore, 1992, p. 681.

35. Olah, G. A., interview in Hargittai, I., *Candid Science: Conversations with Famous Chemists.* Imperial College Press, London, pp. 270–83.

36. Gilman, A. G., interview in Hargittai, I., *Candid Science II: Conversations with Famous Biomedical Scientists.* Imperial College Press, London, 2002, pp. 238–51.

37. Boyer, P. D., interview in Hargittai, I., *Candid Science III: More Conversations with Famous Chemists.* Imperial College Press, London, 2003, pp. 268–79.

38. Hassel, O., in Hargittai, I., *Candid Science: Conversations with Famous Chemists.* Imperial College Press, London, 2000, pp. 158–63.

39. Semenov, N. N., interview in Hargittai, I., *Candid Science: Conversations with Famous Chemists.* Imperial College Press, London, 2000, pp. 466–75.

40. Eigen, M., interview in Hargittai, I., *Candid Science III: More Conversations with Famous Chemists.* Imperial College Press, London, 2003, pp. 368–77.

41. Mössbauer, R., interview in Hargittai, I., *Chem. Intell.*, 1997, 3(3), 6–13.

42. Müller-Hill, B., *Nature*, 1991, 351, 11–12. Herbertz, H.; Müller-Hill, B., *Research Policy*, 1995, 24, 959–79.

43. Marsh, R., conversation in Pasadena, California, 1999, unpublished records.

44. Pauling, L., *Ann. Rev. Biophys. Biophys. Chem.*, 1986, **15**, 1–9.

45. Letter of Linus Pauling to Robert B. Corey, dated 26 February 1968, in the California Institute of Technology Archives. Had the terms Structural Chemistry and Molecular Biology not been capitalized, it could have been taken for a general acknowledgement of Corey's contributions to these fields. However, the postscript most probably refers to the Festschrift for Linus Pauling of the same name, which appeared in 1968 and to which Corey had contributed.

46. Ferry, G., *Dorothy Hodgkin: A Life*. Granta Books, London, 1998, p. 237.

47. Kornberg, A., *For the Love of Enzymes: The Odyssey of a Biochemist*. Harvard University Press, Cambridge, Massachusetts, 1989, p. 172.

48. Karle, J., in *Nobel Lectures: Chemistry 1981–1990*. World Scientific, Singapore, 1992, p. 216.

49. Cornforth, J. W., interview in Hargittai, I., *Candid Science: Conversations with Famous Chemists*. Imperial College Press, London, 2000, pp. 122–37.

50. Brown, H. C., interview in Hargittai, I., *Candid Science: Conversations with Famous Chemists*. Imperial College Press, London, 2000, pp. 250–69.

51. A beautiful example is her letter to the editor, published recently, to set the record straight concerning the steric strain–nonclassical ion controversy. Brown, S. B., *Chem. Intell.*, 1999, **5**(1), 4.

52. Crick, F., private communication, 2000.

53. Caroe, G. M., *William Henry Bragg, 1862–1942: Man and Scientist*. Cambridge University Press, Cambridge and London, 1978. Perutz, M. F., *Nature*, 1978, **276**, 537–39, p. 538.

54. Truter, M. R., *The Guardian*, 3 November 1989 (obituary). See also, Izatt, R. M.; Bradshaw, J. S., *The Oedersen Memorial Issue*. Kluwer Academic, Dordrecht 1992, reprinted from *Journal of Inclusion Phenomena and Molecular Recognition in Chemistry*, 1992, **12**, Nos. 1–4.

55. Levi-Montalcini, R., *In Praise of Imperfection: My Life and Work*. Basic Books, New York, 1988, p. 163.

56. Cohen, S., *Science*, 1975, **187**, 827–30, p. 828.

8 Mentor

1. Gratzer, W., in Watson, J. D., *A Passion for DNA: Genes, Genomes, and Society*. Oxford University Press, 2000, p. 231.

2. Walker, J. E., interview in Hargittai, I., *Candid Science III: More Conversations with Famous Chemists*. Imperial College Press, London, 2003, pp. 280–91.

3. Purcell, E. M., in *Nobel Lectures: Physics 1942–1962*. World Scientific, Singapore, 1998, p. 232.

4. Müller-Hill, B., interview in Hargittai, I., *Candid Science II: Conversations with Famous Biomedical Scientists*. Imperial College Press, London, 2001, pp. 114–29.

5. Levi-Montalcini, R., in *Nobel Lectures: Physiology or Medicine 1981–1990*. Editor-in-charge T. Frängsmyr, editor J. Lindsten. World Scientific, Singapore, 1993, p. 347.

6. Cohen, S., in *Nobel Lectures: Physiology or Medicine 1981–1990*. Editor-in-charge T. Frängsmyr, editor J. Lindsten. World Scientific, Singapore, 1993, pp. 331–2.

7. Tonegawa, S., in *Nobel Lectures: Physiology or Medicine 1981–1990*. Editor-in-charge T. Frängsmyr, editor J. Lindsten. World Scientific, Singapore, 1993, pp. 378–9.

8. Milstein, C., interview in Hargittai, I., *Candid Science II: Conversations with Famous Biomedical Scientists*. Imperial College Press, London, 2002, pp. 220–37.

9. Prelog, V., interview in Hargittai, I., *Candid Science: Conversations with Famous Chemists*. Imperial College Press, London, 2000, pp. 138–47.

10. Eidgenössische Technische Hochschule.

11. Watson, J. D., *A Passion for DNA: Genes, Genomes, and Society*. Oxford University Press, 2000, p. 117.

12. Cram, D. J., interview in Hargittai, I., *Candid Science III: More Conversations with Famous Chemists*. Imperial College Press, London, 2003, pp. 178–97.

13. Klug, A., interview in Hargittai, I., *Candid Science II: Conversations with Famous Biomedical Scientists*. Imperial College Press, London, 2002, pp. 306–29.

14. Caspar, D. L. D.; Klug, A., *Cold Spring Harbor Symp. Quant. Biol.*, 1962, **27**, 1–24.

15. Caroe, G. M., *William Henry Bragg, 1862–1942: Man and Scientist*. Cambridge University Press, Cambridge and London, 1978. Perutz, M. F., *Nature*, 1978, **276**, 537–9, p. 539.

16. Lehn, J.-M., interview in Hargittai, I., *Candid Science III: More Conversations with Famous Chemists*. Imperial College Press, London, 2003, pp.198–205.

17. Schawlow, A., in *Nobel Lectures: Physics 1981–1990*. World Scientific, Singapore, 1993, p. 34.

18. Pais, A., *The Genius of Science: A Portrait Gallery of Twentieth-Century Physicists*. Oxford University Press, 1998, p. 274.

19. Zewail, A. H., interview in Hargittai, I., *Candid Science: Conversations with Famous Chemists*. Imperial College Press, London, 2000, pp. 488–507.

20. Whetten, R. L., interview in Hargittai, I., *Candid Science: Conversations with Famous Chemists*. Imperial College Press, London, 2000, pp. 404–15.

21. Segrè, E. (on the basis of Clarence Larson's interview), Hargittai, I.; Hargittai, M., *Chem. Intell.*, 2000, 6(4), 54–6.

22. Chamberlain, O., conversation in Berkeley, 1999, unpublished records.

23. Jackson, J. D.; Panofsky, W. K. H., *Biographical Memoirs of the National Academy of Sciences of the U. S. A.*, Volume 69. National Academy Press, Washington, DC, 1996, pp. 215–40.

24. Vane, J., interview in Hargittai, I., *Candid Science II: Conversations with Famous Biomedical Scientists*. Imperial College Press, London, 2002, pp. 548–63.

25. Cornforth, J. W., interview in Hargittai, I., *Candid Science: Conversations with Famous Chemists*. Imperial College Press, London, 2000, pp. 122–37.

26. Lederberg, J., interview in Hargittai, I., *Candid Science II: Conversations with Famous Biomedical Scientists*. Imperial College Press, London, 2002, pp. 32–49.

27. *Neurospora* is a fungus, commonly found as a mold on bread that has been kept for too long. It has been much used in biochemical and genetic research.

28. Ottoson, D., in *Nobel Lectures: Physiology or Medicine 1981–1990*. Editor-in-charge T. Frängsmyr, editor J. Lindsten. World Scientific, Singapore, 1993, p. 3.

29. Wiesel, T. N., in *Nobel Lectures: Physiology or Medicine 1981–1990*. Editor-in-charge T. Frängsmyr, editor J. Lindsten. World Scientific, Singapore, 1993, p. 59.

30. Hoffmann, R., interview in *Candid Science: Conversations with Famous Chemists*. Imperial College Press, London, 2000, pp. 190–209.

31. Corey, E. J., in *Nobel Lectures: Chemistry 1981–1990*. Editor-in-charge T. Frängsmyr, editor B. G. Malmström. World Scientific, Singapore, 1992, p. 681.

32. Samuelsson, B., in *Nobel Lectures: Physiology or Medicine 1981–1990.* Editor-in-charge T. Frängsmyr, editor J. Lindsten. World Scientific, Singapore, 1993, p. 117.

33. Blobel, G., interview in Hargittai, I., *Candid Science II: Conversations with Famous Biomedical Scientists.* Imperial College Press, London, 2002, pp. 252–65.

34. Rowland, F. S., interview in Hargittai, I., *Candid Science: Conversations with Famous Chemists.* Imperial College Press, London, 2000, pp. 448–65.

35. Eigen, M., interview in Hargittai, I., *Candid Science III: More Conversations with Famous Chemists.* Imperial College Press, London, 2003, pp. 268–77.

36. Merrifield, B., interview in Hargittai, I., *Candid Science III: More Conversations with Famous Chemists.* Imperial College Press, London, 2003, pp. 206–19.

37. Kornberg, A., interview in Hargittai, I., *Candid Science II: Conversations with Famous Biomedical Scientists.* Imperial College Press, London, 2002, pp. 50–71.

38. Kornberg, A., *For the Love of Enzymes: The Odyssey of a Biochemist.* Harvard University Press, Cambridge, Massachusetts, 1989, p. 172.

39. Berg, P., interview in Hargittai, I., *Candid Science II: Conversations with Famous Biomedical Scientists.* Imperial College Press, London, 2002, pp. 154–81.

40. Perutz, M. F., interview in Hargittai, I., *Candid Science II: Conversations with Famous Biomedical Scientists.* Imperial College Press, London, 2002, pp. 280–95.

41. Gilman, A., interview in Hargittai, I., *Candid Science II: Conversations with Famous Biomedical Scientists.* Imperial College Press, London, 2002, pp. 238–51.

42. Curl, R. F., interview in Hargittai, I., *Candid Science: Conversations with Famous Chemists.* Imperial College Press, London, 2000, pp. 374–84.

43. Glashow, S. L. with Ben Bova, *Interactions: A Journey Through the Mind of a Particle Physicist and the Matter of this World.* Warner Books, New York, 1988, p. 70.

44. Zurer, P., *C&EN*, 27 November 1995, p. 31.

45. Nye, M. J., *Before Big Science: The Pursuit of Modern Chemistry and Physics, 1800–1940.* Harvard University Press, Cambridge, Massachusetts, 1996, p. 196.

46. Bardeen, J., *Proc. R. Soc. Lond. A*, 1980, **371**, 77.

47. Wigner, E. P., *Symmetries and Reflections: Scientific Essays*. Indiana University Press, Bloomington and London, 1967, pp. 262–3.

48. Calvin, M. (on the basis of Clarence Larson's interview), Hargittai, I.; Hargittai, M., *Chem. Intell.*, 2000, 6(1), 52–5.

49. Polanyi, J. C., interview in Hargittai, I., *Candid Science III: More Conversations with Famous Chemists*. Imperial College Press, London, 2003, pp. 378–91

50. Westheimer, F., interview in Hargittai, I., *Candid Science: Conversations with Famous Chemists*. Imperial College Press, London, 2000, pp. 38–53.

9 Changing and combining fields

1. Avery, O. T.; MacLeod, C.; McCarty, M., *J. Exper. Med.*, 1944, **79**, 137–58.

2. Chargaff, E., *Heraclitean Fire: Sketches from a Life before Nature*. Rockefeller University Press, New York, 1978, p. 83.

3. Bernal, J. D., *Biographical Memoirs of Fellows of the Royal Society* (William Astbury), November 1963, 9, 1–35, p. 2.

4. Edelman, G. M., interview in Hargittai, I., *Candid Science II: Conversations with Famous Biomedical Scientists*. Imperial College Press, London, 2002, pp. 196–219.

5. Crick, F., *What Mad Pursuit: A Personal View of Scientific Discovery*. Basic Books, New York, 1988, p. 16.

6. Crick, F., *What Mad Pursuit: A Personal View of Scientific Discovery*. Basic Books, New York, 1988, pp. 74–5.

7. Karle, J., conversation in Washington, 2000, unpublished records.

8. Hauptman, H. A., interview in Hargittai, I., *Candid Science III: More Conversations with Famous Chemists*. Imperial College Press, London, 2003, pp. 292–317.

9. Klug, A., interview in Hargittai, I., *Candid Science II: Conversations with Famous Biomedical Scientists*. Imperial College Press, London, 2002, pp. 306–29.

10. Debye, P. J. W., quoted in Klaw, S., *The New Brahmins: Scientific Life in America*. William Morrow, New York, 1968, p. 127.

11. Rowland, F. S., interview in Hargittai, I., *Candid Science: Conversations with Famous Chemists*. Imperial College Press, London, 2000, pp. 448–65.

12. Lehn, J.-M., interview in Hargittai, I., *Candid Science III: More Conversations with Famous Chemists*. Imperial College Press, London, 2003, pp. 198–205.

13. Fukui, K., interview in Hargittai, I., *Candid Science: Conversations with Famous Chemists*. Imperial College Press, London, 2000, pp. 210–21.

14. Eigen, M., interview in Hargittai, I., *Candid Science III: More Conversations with Famous Chemists*. Imperial College Press, London, 2003, pp. 268–77.

15. Sanger, F., interview in Hargittai, I., *Candid Science II: Conversations with Famous Biomedical Scientists*. Imperial College Press, London, 2002, pp. 72–83.

16. Herring, C., *Physics Today*, April 1992, 26–33, p. 32.

17. Berg, P., interview in Hargittai, I., *Candid Science II: Conversations with Famous Biomedical Scientists*. Imperial College Press, London, 2002, pp. 154–81.

18. Szent-Györgyi, A., *Ann. Rev. Biochem.*, 1963, 32, 1–14, p. 9.

19. Hargittai, I., *Chem. Intell.*, 1999, 5(1), 43–9, p. 46.

20. Alvarez, W., *T. Rex and the Crater of Doom*. Princeton University Press, 1997.

21. Kroto, H. W., interview in Hargittai, I., *Candid Science: Conversations with Famous Chemists*. Imperial College Press, London, 2000, pp. 332–57.

22. Smalley, R. E., interview in Hargittai, I., *Candid Science: Conversations with Famous Chemists*. Imperial College Press, London, 2000, pp. 362–73.

23. Curl, R. F., interview in Hargittai, I., *Candid Science: Conversations with Famous Chemists*. Imperial College Press, London, 2000, pp. 374–87.

24. Porter, G., interview in Hargittai, I., *Candid Science: Conversations with Famous Chemists*. Imperial College Press, London, 2000, pp. 476–84.

25. Elion, G. B., in *Nobel Lectures: Physiology or Medicine 1981–1990*. Editor-in-charge T. Frängsmyr, editor J. Lindsten. World Scientific, Singapore, 1993, pp. 444–5.

26. Prigogine, I., interview in Hargittai, I., *Candid Science III: More Conversations with Famous Chemists*. Imperial College Press, London, 2003, pp. 422–31.

27. Pauling, L., *J. Am. Chem. Soc.*, 1940, 62, 2643–53. Pauling, L., *Ann. Rev. Biophys. Biophys. Chem.*, 1986, 15, 1–9.

28. Mullis, K. B., interview in Hargittai, I., *Candid Science II: Conversations with Famous Biomedical Scientists*. Imperial College Press, London, 2002, pp. 182–95.

29. Mullis, K. B., *Nature*, 1968, **218**, 663–4.

30. Altman, S., interview in Hargittai, I., *Candid Science II: Conversations with Famous Biomedical Scientists*. Imperial College Press, London, 2002, pp. 338–49.

31. Pople, J. A., interview in Hargittai, I., *Candid Science: Conversations with Famous Chemists*. Imperial College Press, London, 2000, pp. 178–89.

32. Lawton, G., *Chemistry & Industry*, 2 November 1998, p. 873.

33. Jackson, J. D.; Panofsky, W. K. H., *Biographical Memoirs of the National Academy of Sciences of the U. S. A.*, Vol. 69. National Academy Press, Washington, DC, 1996, pp. 215–40 (Edwin McMillan).

34. Gilbert, W., interview in Hargittai, I., *Candid Science II: Conversations with Famous Biomedical Scientists*. Imperial College Press, London, 2002, pp. 98–113.

35. Chargaff, E., *Heraclitean Fire: Sketches from a Life before Nature*. Rockefeller University Press, New York, 1978.

36. Chargaff, E., *Heraclitean Fire: Sketches from a Life before Nature*. Rockefeller University Press, New York, 1978, p. 14. For a conversation with Erwin Chargaff and a sample of quotations from his books, see Hargittai, I., *Candid Science: Conversations with Famous Chemists*. Imperial College Press, London, 2000, pp. 14–37.

10 Making an impact

1. Klug, A., interview in Hargittai, I., *Candid Science II: Conversations with Famous Biomedical Scientists*. Imperial College Press, London, 2002, pp. 306–29.

2. Kroto, H. W., interview in Hargittai, I., *Candid Science: Conversations with Famous Chemists*. Imperial College Press, London, 2000, pp. 332–57.

3. Seaborg, G. T., interview in Hargittai, I., *Candid Science III: More Conversations with Famous Chemists*. Imperial College Press, London, 2003, pp. 2–17.

4. Olah, G. A., *J. Am. Chem. Soc.*, 1972, **94**, 808–20.

5. Olah, G. A., in *Nobel Lectures: Chemistry 1991–1995*. Edited by B. G. Malmström, World Scientific, Singapore, 1997, p. 161.

6. Eigen, M., interview in Hargittai, I., *Candid Science III: More Conversations with Famous Chemists*. Imperial College Press, London, 2003, pp. 268–77.

7. *Reflections on W. A. Engelhardt* (in Russian). Nauka, Moscow, 1989. (In modern transliteration it is V. A. Engelhardt.)

8. Engelhardt, W. A.; Lyubimova, M. N., *Nature*, 1939, **144**, 668–9.

9. This was the meeting where Marshall Nirenberg made his historic report on cracking the genetic code (see, p. 98).

10. Engelhardt, W. A., *Ann. Rev. Biochem.* 1982, **51**, 1.

11. Weaver, W., *Science*, 1970, **170**, 581–2.

12. Rees, M., *Biographical Memoirs of the National Academy of Sciences of the U. S. A.*, Vol. 57. National Academy Press, Washington, DC, 1987, pp. 493–530, p. 503 (Warren Weaver).

13. Mullis, K. B., interview in Hargittai, I., *Candid Science II: Conversations with Famous Biomedical Scientists*. Imperial College Press, London, 2002, pp. 182–95.

14. Luria, S. E., *BioScience*, 1970, **20**, 1289–93, 1296, p. 1289.

15. Luria, S. E., *BioScience*, 1970, **20**, 1289–93, 1296, p. 1292.

16. Truter, M. R., *The Guardian*, 3 November 1989 (obituary of Charles Pedersen).

17. Cram, D. J., interview in Hargittai, I., *Candid Science III: More Conversations with Famous Chemists*. Imperial College Press, London, 2003, pp. 178–97.

18. See, for example, Furukawa, Y., *Inventing Polymer Science: Staudinger, Carothers, and the Emergence of Macromolecular Chemistry*. University of Pennsylvania Press, Philadelphia, 1998.

19. Serber, R. with Crease, R. P., *Peace & War: Reminiscences of a Life on the Frontiers of Science*. Columbia University Press, New York, 1998, p. 200.

20. Gell-Mann, M., *The Quark and the Jaguar: Adventures in the Simple and the Complex*. W. H. Freeman, New York, 1994, p. 11.

21. The similarity between the shapes of the atomic nucleus and a drop of liquid water was first suggested by George Gamow and the idea was then extended by Niels Bohr. See Wheeler, J. A. with Ford, K., *Geons, Black Holes, and Quantum Foam: A Life in Physics*. W. W. Norton, New York, 2000, p. 20.

22. Frisch, O. R., *What Little I Remember*. Cambridge University Press, 1979.

23. Wheeler, J. A. with Ford, K., *Geons, Black Holes, and Quantum Foam: A Life in Physics*. W. W. Norton, New York, 2000, pp. 21–2.

24. Bohr, N.; Wheeler, J. A., *Phys. Rev.*, 1939, **56**, 426–50.

25. Maddox, J., interview in Hargittai, I., *Chem. Intell.*, 1999, **5**(2), 53–5.

26. Mullis, K.; Faloona, F.; Scharf, S.; Saiki, R.; Horn, G.; Erlich, H., *Cold Spring Harbor Symp. Quant. Biol.*, 1986, **51**, 260.

27. Michel, H., *J. Mol. Biol.*, 1982, **158**, 567–72.

28. Deisenhofer, J.; Epp, O.; Miki, K.; Huber, R.; Michel, H., *Nature*, 1985, **318**, 618–24.

29. Gratzer, W., private communication, 2001.

30. Perutz, M. F., *I Wish I'd Made You Angry Earlier*. Oxford University Press, 1998, pp. 297–8.

31. Boyer, P. D., interview in Hargittai, I., *Candid Science III: More Conversations with Famous Chemists*. Imperial College Press, London, 2003, pp. 268–79.

32. Edelman, G. M., interview in Hargittai, I., *Candid Science II: Conversations with Famous Biomedical Scientists*. Imperial College Press, London, 2002, pp. 196–219.

33. Edelman, G. M., *J. Am. Chem. Soc.*, 1959, **81**, 3155–6.

34. Prigogine, I., interview in Hargittai, I., *Candid Science III: More Conversations with Famous Chemists*. Imperial College Press, London, 2003, pp. 422–31.

35. Belousov, B. P., *Sb. Ref. Radiats. Med. za 1958*, Medgiz, Moscow, 1959, **1**, 145 (in Russian).

36. Zhabotinsky, A. M., interview in Hargittai, I., *Candid Science III: More Conversations with Famous Chemists*. Imperial College Press, London, 2003, pp. 432–47.

37. Zhabotinsky, A. M., *Biofizika*, 1964, **9**, 306 (in Russian).

38. Shechtman, D.; Blech, I.; Gratias, D.; Cahn, J. W., *Phys. Rev. Lett.*, 1984, **53**, 1951–3.

39. Bastiansen, O.; Hassel, O., *Nature*, 1946, **157**, 765.

40. Hassel, O., *Tidsskrift for kjemi, bergvesen og metallurgi*, 1943, **5**, 32–4 (in Norwegian).

41. Levi-Montalcini, R., *In Praise of Imperfection: My Life and Work*. Basic Books, New York, 1988, p. 84.

42. Molina, M. J.; Rowland, F. S., *Nature*, 1974, **249**, 810–12.

43. Kornberg, A., interview in Hargittai, I., *Candid Science II: Conversations with Famous Biomedical Scientists*. Imperial College Press, London, 2002, pp. 50–71.

44. Nirenberg, M. W., profile in Hargittai, I., *Candid Science II: Conversations with Famous Biomedical Scientists*. Imperial College Press, London, 2002, pp. 130–41.

45. Altman, S., interview in Hargittai, I., *Candid Science II: Conversations with Famous Biomedical Scientists*. Imperial College Press, London, 2002, pp. 338–49.

46. Yalow, R., in *Nobel Lectures: Physiology or Medicine 1971–1980*. Edited by J. Lindsten. World Scientific, Singapore, 1992, pp. 448–9.

47. Blobel, G., interview in Hargittai, I., *Candid Science II: Conversations with Famous Biomedical Scientists*. Imperial College Press, London, 2002, pp. 252–65.

48. Olah, G. A., interview in Hargittai, I., *Candid Science: Conversations with Famous Chemists*. Imperial College Press, London, 2000, pp. 270–83.

49. For a summary, see Olah, G. A., in *Nobel Lectures: Chemistry 1991–1995*. Edited by B. G. Malmström. World Scientific, Singapore, 1997, pp. 162–5.

50. Gronowitz, S., in *Nobel Lectures: Chemistry 1991–1995*. Edited by B. G. Malmström. World Scientific, Singapore, 1997, p. 139.

51. Perutz, M., *Acta Cryst.*, 1970, A **26**, 183–5, p. 184.

52. Cornforth, J. W., interview in Hargittai, I., *Candid Science: Conversations with Famous Chemists*. Imperial College Press, London, 2000, pp. 122–37.

53. Lowry, O. H.; Rosebrough, N. J.; Farr, A. L.; Randall, R. J., 'Protein measurement with the Folin phenol reagent'. *J. Biol. Chem.*, 1951, **193**, 265–76.

54. Nathans, D., interview in Hargittai, I., *Candid Science II: Conversations with Famous Biomedical Scientists*. Imperial College Press, London, 2002, pp. 142–53.

11 *Is there life after the Nobel prize?*

1. Robbins, F. C., interview by M. Hargittai in Hargittai, I., *Candid Science II: Conversations with Famous Biomedical Scientists*. Imperial College Press, London, 2002, pp. 498–517.

2. Mullis, K. B., *Dancing Naked in the Mind Field*. Pantheon Books, New York, 1998, p. 21.

3. Gilman, A., interview in Hargittai, I., *Candid Science II: Conversations with Famous Biomedical Scientists*. Imperial College Press, London, 2002, pp. 238–51.

4. Pais, A., *The Genius of Science: A Portrait Gallery*. Oxford University Press, 2000, p. 273.

5. Bernstein, J., *The New Yorker*, 13 and 20 October 1975, as quoted by Pais, A., *The Genius of Science*, p. 273.

6. M. Hargittai's conversation with Gösta Ekspong in Stockholm, 2001, unpublished records.

7. Schwartz, M., in *Nobel Lectures: Physics 1981–1990*. Editor-in-charge T. Frängsmyr, editor G. Ekspong. World Scientific, Singapore, 1993, p. 466.

8. Mössbauer, R., interview in Hargittai, I., *Chem. Intell.*, 1997, 3(3), 6–19.

9. I am grateful to Anders Bárány for the story of his grandfather, Robert Bárány, Stockholm, 2000.

10. Karle, J., in *Nobel Lectures: Chemistry 1981–1990*. Editor-in-charge T. Frängsmyr, editor B. G. Malmström, World Scientific, Singapore, 1992, p. 217.

11. Chamberlain, O., conversation in Berkeley, California, May 2000, unpublished records.

12. Brown, H. C., interview in Hargittai, I., *Candid Science: Conversations with Famous Chemists*. Imperial College Press, London, 2000, pp. 250–69.

13. Sanger, F., interview in Hargittai, I., *Candid Science II: Conversations with Famous Biomedical Scientists*. Imperial College Press, London, 2002, pp. 72–83.

14. Milstein, C., interview in Hargittai, I., *Candid Science II: Conversations with Famous Biomedical Scientists*. Imperial College Press, London, 2002, pp. 220–37.

15. Bignami, G., *Sapere*, 2000, 66(5), 60–5 (in Italian).

16. Vane, J. R., interview in Hargittai, I., *Candid Science II: Conversations with Famous Biomedical Scientists*. Imperial College Press, London, 2002, pp. 548–63.

17. Mullis, K. B., interview in Hargittai, I., *Candid Science II: Conversations with Famous Biomedical Scientists*. Imperial College Press, London, 2002, pp. 182–95.

18. Elion, G. B., interview in Hargittai, I., *Candid Science: Conversations with Famous Chemists*. Imperial College Press, London, 2000, pp. 54–71.

19. Merrifield, B., interview in Hargittai, I., *Candid Science III: More Conversations with Famous Chemists*. Imperial College Press, London, 2003, pp. 206–19.

20. Yalow, R., interview in Hargittai, I., *Candid Science II: Conversations with Famous Biomedical Scientists*. Imperial College Press, London, 2002, pp. 518–23.

21. Straus, E., *Rosalyn Yalow: Nobel Laureate, Her Life and Work in Medicine, A Biographical Memoir*. Plenum, New York and London, 1998, p. 23.

22. Baum, R., *C&EN*, 30 January 1995, p. 34.

23. Private communication from George Olah, February 1996.

24. Hargittai, I., *Chem. Intell.*, 1997, 3(1), p. 24.

25. Conversations with Professor Emeritus Ferenc Guba of Szeged University and Professor Emeritus Andrew Szent-Györgyi of Brandeis University, in Budapest, 1999, unpublished records.

26. Conversation with Rita Levi-Montalcini, by M. Hargittai, Rome, 2000, in Hargittai, I., *Candid Science II: Conversations with Famous Biomedical Scientists*. Imperial College Press, London, 2002, pp. 364–75.

27. Calvin, M. (based on Clarence Larson's interview), in Hargittai, I.; Hargittai, M., *Chem. Intell.*, 2000, 6(1), 52–5.

28. Cram, D. J., interview in Hargittai, I., *Candid Science III: More Conversations with Famous Chemists*. Imperial College Press, London, 2003, pp. 178–97.

29. Kornberg, A., interview in Hargittai, I., *Candid Science II: Conversations with Famous Biomedical Scientists*. Imperial College Press, London, 2002, pp. 50–71.

30. Black, J. W., profile in Hargittai, I., *Candid Science II: Conversations with Famous Biomedical Scientists*. Imperial College Press, London, 2002, pp. 524–41.

31. Rouhi, M., *C&EN*, 23 March 1998, p. 10.

32. Roberts, J. D., *Chem. Intell.*, 1999, 5(1), p. 32.

33. Curl, R. F., interview in Hargittai, I., *Candid Science: Conversations with Famous Chemists*. Imperial College Press, London, 2000, pp. 374–87.

34. Fukui, K., interview in Hargittai, I., *Candid Science: Conversations with Famous Chemists*. Imperial College Press, London, 2000, pp. 210–21.

35. Morse, P., *C&EN*, 1 June 1998, p. 48.

36. Truter, M. R., *The Guardian*, 3 November 1989 (obituary of Charles Pedersen).

37. Nirenberg, M. W., profile in Hargittai, I., *Candid Science II: Conversations with Famous Biomedical Scientists*. Imperial College Press, London, 2002, pp. 130–41.

38. The neutrino is a particle that carries no charge and which is subject only to weak interactions. Gravitation alone is weaker, very much weaker, and it matters only on astronomical scales. On nuclear scales, gravitation is negligibly small; this is so even when it is compared to the so-called weak interaction, which is so weak that a neutrino entering the Sun on one side and leaving it on the other hardly feels its existence. Neutrinos, in particular, are formed inside the Sun, as they are in other stars, when four protons are fused into a helium nucleus. The helium nucleus is a little lighter than the four protons and this mass difference is converted into energy, which the Sun radiates. The generated neutrinos then leave the Sun, some coming to the Earth where we can measure a tiny fraction of them. On the GALLEX project, see, for example, Hargittai, M.; Hargittai, I., *Chem. Intell.*, 1997, 3(3), 10–11.

39. Eigen, M., interview in Hargittai, *Candid Science III: More Conversations with Famous Chemists.* Imperial College Press, London, 2003, pp. 368–77.

40. He did not abandon his first love, music, and made a CD on which he played Mozart, beautifully, accompanied by the Chamber Orchestra, Basel, conducted by Paul Sacher in 1981, and by the New Orchestra of Boston, conducted by David Epstein in 1991. Eigen's professional-level music is not unique among Nobel laureates. Max Planck could play the piano with professional skill, and he, violinist Joseph Joachim, and violinist and fellow physicist Albert Einstein (P21), played as a trio. Planck composed songs and an operetta in his youth, served as a choir master, and conducted an orchestra. Robert Root-Bernstein tabulated the artistic proclivities of famous scientists, including quite a few Nobel laureates. There are many painters, musicians, poets, and writers among them. See Heilbron, J. L., *The Dilemmas of an Upright Man: Max Planck as Spokesman for German Science.* University of California Press, Berkeley, 1986, p. 34. Root-Bernstein, R. S., *Discovering.* Harvard University Press, Cambridge, Massachusetts and London, 1991, pp. 318–27.

41. Ernst, R. R., interview in Hargittai, I., *Candid Science: Conversations with Famous Chemists.* Imperial College Press, London, 2000, pp. 294–307.

42. Lubkin, G. B., *Physics Today*, November 1992, 34–40, p. 40.

43. Layman, P., *C&EN*, 13 July 1998, pp. 33, 34.

44. Klug, A., interview in Hargittai, I., *Candid Science II: Conversations with Famous Biomedical Scientists.* Imperial College Press, London, 2002, pp. 306–29.

45. Rowland, F. S., interview in Hargittai, I., *Candid Science: Conversations with Famous Chemists*. Imperial College Press, London, 2000, pp. 448–65.

46. Crutzen, P. J., interview in Hargittai, I., *Candid Science III: More Conversations with Famous Chemists*. Imperial College Press, London, 2003, pp. 460–5.

47. Watson, J. D., interview in Hargittai, I., *Candid Science II: Conversations with Famous Biomedical Scientists*. Imperial College Press, London, 2002, pp. 2–15.

48. Berg, P., interview in Hargittai, I., *Candid Science II: Conversations with Famous Biomedical Scientists*. Imperial College Press, London, 2002, pp. 154–81.

49. Seaborg, G. T., *A Chemist in the White House: From the Manhattan Project to the End of the Cold War*. American Chemical Society, Washington, DC, 1998.

50. Glashow, S. with Bova, B., *Interactions: A Journey Through the Mind of a Particle Physicist and the Matter of this World*. Warner Books, New York, 1988, p. 42.

51. Polanyi, J. C., interview in Hargittai, I., *Candid Science III: More Conversations with Famous Chemists*. Imperial College Press, London, 2003, pp. 378–91.

52. Hoffmann, R., interview in Hargittai, I., *Candid Science: Conversations with Famous Chemists*. Imperial College Press, London, 2000, pp. 190–209.

53. Porter, G., interview in Hargittai, I., *Candid Science: Conversations with Famous Chemists*. Imperial College Press, London, 2000, pp. 476–87.

54. Lederman, L. with Teresi, D., *The God Particle: If the Universe Is the Answer, What Is the Question?* Delta, New York, 1994.

55. Lederman, L., interview in Hargittai, I., *Chem. Intell.*, 1998, 4(4), 20–9.

56. Wilson, K. G., conversation in Columbus, Ohio, 2000, unpublished records.

57. Gilman, A. G., interview in Hargittai, I., *Candid Science II: Conversations with Famous Biomedical Scientists*. Imperial College Press, London, 2002, pp. 238–51.

58. This is extracted from Seltzer, R., *C&EN*, 24 April 1995, p. 52.

59. Hoffmann, R.; Schmidt, S. L., *Old Wine, New Flasks: Reflections on Science and Jewish Tradition*. W. H. Freeman, New York, 1997.

60. Anderson, P. W., interview in Hargittai, I., *Chem. Intell.*, 2000, 6(3), 28–32.

61. Pais, A., *The Genius of Science: A Portrait Gallery*. Oxford University Press, 2000, p. 346.

62. Wigner, E. P., conversations in Austin, Texas, 1969.

63. Pais, A., *The Genius of Science: A Portrait Gallery*. Oxford University Press, 2000, p. 347.

64. Klein, E., private communication, Stockholm, 2000.

65. Dyson, F., conversation with M. Hargittai, Princeton, 2000, unpublished records.

66. *Time*, 16 October 2000, p. 73.

67. Josephson, B., conversation in Cambridge, 2000, unpublished records.

68. *Time*, 29 March 1999, p. 164.

69. Straus, E., *Rosalyn Yalow, Nobel Laureate: Her Life and Work in Medicine, A Biographical Memoir*. Plenum, New York and London, 1998, p. 244.

70. I attended Robert J. Paradowski's (Rochester Institute of Technology) talk about Linus Pauling at a Pauling symposium during the spring 1995 meeting of the American Chemical Society in Anaheim. There is a summary of his talk in Borman, S., *C&EN*, 8 May 1995, p. 31.

71. My wife and I saw the play in London in 2000.

72. Schwarzchild, B., *Physics Today*, May 2000, pp. 51–2.

73. Cassidy, D. C., *Uncertainty: The Life and Science of Werner Heisenberg*. W. H. Freeman, New York, 1992.

74. Heisenberg must have been aware of the Nazi crimes. He visited his old friend Hans Frank, the 'Butcher of Poland', in Poland in 1943, and he also negotiated the accelerated production of uranium plates for his experiments using slave labor. See, Rose, P. L., *Heisenberg and the Nazi Atomic Bomb Project: A Study of German Culture*. University of California Press, Berkeley, 1998, pp. 284–5, 309.

12 *Who did not win*

1. Merrifield, B., interview in Hargittai, I., *Candid Science III: More Conversations with Famous Chemists*. Imperial College Press, London, 2003, pp. 206–19.

2. Garfield, E., *Current Contents*, 1986, No. 23 (9 June), p. 8.

3. Maddox, J., *The Independent*, 11 October 2000, p. 5.

4. Zuckerman, H., *Scientific Elite: Nobel Laureates in the United States.* Transaction Publishers, New Brunswick, New Jersey, and London, 1996.

5. *Nobel: The Man & His Prizes.* Third Edition. Edited by the Nobel Foundation and W. Odelberg. American Elsevier, New York, 1972, p. 199.

6. *Nobel: The Man & His Prizes.* Third Edition. Edited by the Nobel Foundation and W. Odelberg. American Elsevier, New York, 1972, p. 201.

7. *Nobel: The Man & His Prizes.* Third Edition. Edited by the Nobel Foundation and W. Odelberg. American Elsevier, New York, 1972, p. 225.

8. Hodgkin, D. C.; Riley, D. P., in *Structural Chemistry and Molecular Biology.* Edited by A. Rich and N. Davidson. W. H. Freeman, San Francisco and London, 1968, pp. 15–28.

9. Perutz, M., *London Review of Books*, 6 July 2000, p. 35, in a book review of *J. D. Bernal, A Life in Science and Politics.* Edited by B. Swann and F. Aprahamian, Verso, London, 1999.

10. Perutz, M. F., *New Scientist*, 21 October 1976, 144–7, p. 144.

11. M. Goldhaber, letter of nomination, 14 December 1948, Nobel Archives.

12. Watson, J. D., interview in Hargittai, I., *Candid Science II: Conversations with Famous Biomedical Scientists.* Imperial College Press, London, 2002, pp. 2–15.

13. Jacob, F., interview in Hargittai, I., *Candid Science II: Conversations with Famous Biomedical Scientists.* Imperial College Press, London, 2002, pp. 84–97.

14. Novick, A., in *Phage and the Origin of Molecular Biology.* Edited by J. Cairns, G. Stent, and J. D. Watson. Cold Spring Harbor Laboratory of Quantitative Biology, 1966, p. 134.

15. The Hungarian word 'szilárd' means solid. The accent disappeared from Leo Szilard's name.

16. See, for example, Kornberg, A., *For the Love of Enzymes.* Harvard University Press, Cambridge, Massachusetts, and London, 1989, p. 136.

17. Sanger, F., interview in Hargittai, I., *Candid Science II: Conversations with Famous Biomedical Scientists.* Imperial College Press, London, 2002, pp. 72–83.

18. The genetics of bacteria was yet to come in Joshua's Lederberg's (M58) work.

19. Dale, H., Anniversary Address to the Royal Society, 1945, pp. 1–17, p. 2.

20. Oswald Avery was nominated in 1932 by 5 nominators, in 1933 by 1, 1934/2, 1936/2, 1937/1, 1938/2, 1939/2, 1942/1, 1946/2, 1947/3, 1948/4, 1949/4, and 1950/1; Archives of the Karolinska Institute.

21. Dubos, R. J., *The Professor, The Institute, and DNA*. Rockefeller University Press, New York, 1976, p. 159.

22. 'His crown of glory was complete; ours lacked only him.' (Translation by Alan L. Mackay, FRS, London, private communication, 2001.)

23. P. W. Bridgman, letter of nomination, 15 January 1947, Nobel Archives.

24. I. I. Rabi, letter of nomination, 17 January 1947, Nobel Archives.

25. Ingmar Bergström, private communication, Budapest, 2001.

26. Letters of nomination, 1949, Nobel Archives.

27. C. K. Ingold, letter of nomination, 7 January 1940, Nobel Archives.

28. Lipscomb, W. N., interview in Hargittai, I., *Chem. Intell.*, 1996, 2(3), 6–11.

29. G. I. Finch, undated letter of nomination, received 27 January 1940, Nobel Archives.

30. Landsteiner left Vienna in 1919 and eventually joined the Rockefeller Institute in New York in 1922. His Nobel award was 'for his discovery of human blood groups'.

31. Enders, J. F.; Weller, T. H.; Robbins, F. C., *Science*, 1949, 109, 85.

32. Robbins, F. C.; Daniel, T. M., in *Polio*. Edited by T. M. Daniel and F. C. Robbins. University of Rochester Press, 1997, pp. 5–22.

33. Watson, J. D., private communication, 2000.

34. Herzberg, G., *Ann. Rev. Phys. Chem.*, 1985, 36, 1–30.

35. Teller, E., interview in Hargittai, I.; Hargittai, M., *Chem. Intell.*, 1997, 3(1), 14–23.

36. The Jahn–Teller effect describes the symmetry lowering of molecules in a degenerate electronic state to an energetically more advantageous structure.

37. Bartlett, N., interview in Hargittai, I., *Candid Science III: More Conversations with Famous Chemists*. Imperial College Press, London, 2003, pp. 28–47.

38. Bartlett, N., *Proc. Chem. Soc.*, 1962, 218.

39. Yost, D. M., in *Noble-Gas Compounds*. Edited by H. H. Hyman, University of Chicago Press, 1963, pp. 21–2.

40. Levi, P., *The Periodic Table*. Translated from the Italian by
    R. Rosenthal, Schocken Books, New York, 1984, p. 3.

41. Pyykkö, P., *Science*, 2000, **290**, 64–5. Evans, C. J.; Lessari, A. A.;
    Gerry, M. C. L., *J. Amer. Chem. Soc.*, 2000, **122**, 6100–5. *Phys. Chem.
    Chem. Phys.*, 2000, **2**, 3943–8.

42. Ernster, L., interview in Hargittai, I., *Candid Science II: Conversations
    with Famous Biomedical Scientists*. Imperial College Press, London,
    2002, pp. 376–95.

43. Perutz, M., *Selected Topics in the History of Biochemistry: Personal
    Recollections, V*. Edited by G. Semenza and R. Jaenicke. Elsevier,
    Amsterdam, 1997, pp. 57–67, p. 65.

44. David Keilin received nominations in 1932 from one nominator, 1937
    from two, 1938 from three, and from one each year in 1939, 1940,
    1941, 1945, 1946, and 1950; Archives of the Karolinska Institute.

45. Kamen, M. D., *Radiant Science, Dark Politics: A Memoir of the
    Nuclear Age*. University of California Press, Berkeley, 1985.

46. Libby could have also shared the prize with his former PhD student,
    E. C. Anderson, who in his doctoral studies developed a technique
    for measuring low levels of radioactivity from living material (Hedges,
    R. E. M., *Education in Chemistry*, November 1997, pp. 157–9, 164).
    Their joint work was the dating method applicable to biological
    remains. However, Libby received the Nobel Prize alone. It
    was usual at that time to omit doctoral students and junior associates
    from the prize, a practice which seems to have changed lately.

47. *Physics Today*, February 1996, p. 73.

48. *Fifty Years of X-ray Diffraction: Dedicated to the International Union
    of Crystallography on the Occasion of the Commemoration Meeting in
    Munich, July 1962*. Edited by P. P. Ewald, with contributions from
    numerous crystallographers. A. Oosthoek's Uitversmaatschappij NV,
    Utrecht, The Netherlands, 1962.

49. Crawford, E.; Sime R. L.; Walker, M., *Physics Today*, September
    1997, pp. 26–32.

50. Sime, R. L., *Lise Meitner: A Life in Physics*. University of California
    Press, Berkeley, 1996.

51. Bergström, I., *Årsberättelse 1999, Kungl. Vetenskapsakademiens* (in
    Swedish, Yearbook 1999 of the Royal Swedish Academy of Sciences),
    1999, pp. 17–25. I thank professor Bergström for the English
    translation of the published version of his lecture and also of the
    original version. The start of Ingmar Bergström's graduate studies
    coincided with the last year of Lise Meitner's stay in Manne
    Siegbahn's institute in Stockholm. Bergström was Siegbahn's student

and later his successor as director of what is today the Manne Siegbahn Institute of Physics of Stockholm University. (Unpublished records of conversations with Ingmar Bergström in Budapest, 2001.) I am also grateful to Professor Bergström for donating to me the complete documentation he had compiled in the course of his preparation for the Meitner Lecture.

52. The first female member was elected in the eighteenth century so Lise Meitner was the first female member in modern science.

53. Max von Laue, letter of nomination, dated December 1945 (received 28 January 1946), Nobel Archives.

54. Niels Bohr, letter of nomination, 21 January 1946, Nobel Archives.

55. Oskar Klein, letter of nomination, 30 January 1946, Nobel Archives.

56. E. A. Hylleraas (Oslo), letter of nomination, 29 January 1946, Nobel Archives.

57. James Franck, letter of nomination, 5 January 1945 (received 9 February), Nobel Archives.

58. Kasimir Fajans, letter of nomination, 9 January 1946, Nobel Archives.

59. A. H. Compton, letter of nomination, 10 January 1947, Nobel Archives. Compton's detailed letter is accompanied by support material.

60. M. Planck, letter of nomination, 14 January 1947, Nobel Archives. Planck, shortly before his death, writes only one sentence: 'Für die Verleihung des dieses jährigen Nobelpreises für Physik schlage ich vor: Frau Professor Lise Meitner, Stockholm.'

61. M. de Broglie, letter of nomination, 22 January 1947, Nobel Archives.

62. O. Klein, letter of nomination, 27 January 1947, Nobel Archives.

63. E. A. Hyleraas, letter of nomination, 25 January 1947, Nobel Archives.

64. L. de Broglie, letter of nomination, 22 January 1947, Nobel Archives.

65. N. Bohr, letter of nomination, 29 January 1947, Nobel Archives.

66. N. R. Dhar, letter of nomination, 12 November 1946, Nobel Archives.

67. Harald Wergeland (Trondheim), letter of nomination, 28 January 1948, Nobel Archives.

68. Otto Hahn, letter of nomination, 30 January 1948, Nobel Archives.

69. O. Klein, letter of nomination, 30 January 1948, Nobel Archives.

70. Gerhard Hettner (Munich), letter of nomination, 24 January 1948 (received 26 February), Nobel Archives.

71. Conversation with Anders Bárány, Stockholm, 2000, unpublished records.

72. Crawford, E.; Sime, R. L.; Walker, M., *Nature* 1996, **382**, 393.

73. I. V. Stalin is supposed (fictitiously) to have said, 'The following are the minutes of the next meeting.' (By Alan L. Mackay, FRS, London.)

74. Garwin, R. L.; Lee, T.-D., *Physics Today*, October 1997, p. 122.

75. Lederman, L., interview in Hargittai, I., *Chem. Intell.*, 1998, 4(4), 20–9.

76. Segrè, E. (based on Clarence Larson's interview), in Hargittai, I.; Hargittai, M., *Chem. Intell.*, 2000, 6(4), 54–6.

77. Chamberlain, O., conversation in Berkeley, 2000, unpublished records.

78. Heilbron, J. L., *Physics Today*, November 1992, 42–7, p. 46.

79. Klug, A., interview in Hargittai, I., *Candid Science II: Conversations with Famous Biomedical Scientists*. Imperial College Press, London, 2002, pp. 306–29.

80. Bartholomew, J. R., *Osiris*, 1998, 13, 238–84, p. 280.

81. Rechenberg, H., *Physics Today*, October 1997, pp. 126–7.

82. Mulliken, R. S., in *Nobel Lectures: Chemistry 1963–1970*. World Scientific, Singapore, 1999, p. 141.

83. Pitzer, K. S., interview in Hargittai, I., *Candid Science: Conversations with Famous Chemists*. Imperial College Press, London, 2000, pp. 438–47.

84. Hewish, A., conversation in Cambridge, 2000, unpublished records.

85. Bell, J., conversation with M. Hargittai in Princeton, 2000, unpublished records.

86. Quoted by Freeman Dyson in a conversation with M. Hargittai, Princeton, 2000, unpublished records.

87. Yalow, R. S., interview in Hargittai, I., *Candid Science II: Conversations with Famous Biomedical Scientists*. Imperial College Press, London, 2002, pp. 518–23.

88. Straus, E., *Rosalyn Yalow: Nobel Laureate. Her Life and Work in Medicine, A Biographical Memoir*. Plenum, New York and London, 1998, p. 227.

89. Wilson, K., conversation in Columbus, Ohio, 2000, unpublished records.

90. Rowland, F. S., interview in Hargittai, I., *Candid Science: Conversations with Famous Chemists*. Imperial College Press, London, 2000, pp. 448–65.

91. *Physics Today*, February 1999, p. 80.

92. Krätschmer, W.; Lamb, L. D.; Fostiropoulos, K.; Huffman, D. R., 'Solid C_{60}: a new form of carbon'. *Nature*, 1990, **347**, 354–8.

93. Curl, R. F., interview in Hargittai, I., *Candid Science: Conversations with Famous Chemists.* Imperial College Press, London, 2000, pp. 374–87.

94. Furchgott, R. F., interview in Hargittai, I., *Candid Science II: Conversations with Famous Biomedical Scientists.* Imperial College Press, London, 2002, pp. 578–93.

95. See, for example, the commentary in *Science*, 1996, **274**, 173–4.

96. See, for example, *Nature*, 1998, **395**, 625–6.

97. Palmer, R. M. J.; Ferrige, A. G.; Moncada, S., *Nature*, 1987, **327**, 524–6.

98. Helmuth, L., *Science*, 2001, **291**, 567, 569.

99. Karle, J., conversation in Washington, DC, 2000, unpublished records.

Epilogue

1. Paraphrasing 'If God did not exist, it would be necessary to invent him' by Voltaire (epistle to the author of *Livres des trois Imposteurs*). See, for example, *The Pocket Book of Quotations*. Edited by H. Davidoff, Pocket Books, New York, 1942, p. 118.

2. Furberg, S., conversations in Oslo, 1970s. Sven Furberg was another of Odd Hassel's students. He made an important contribution to uncovering the structure of DNA and is cited in Watson and Crick's original report.

3. Hargittai, I., *Chem. Intell.*, 2000, **6**(2), 37–40.

Acknowledgements

1. Published by Imperial College Press, London.

Further reading

The following publications are closely related to the Nobel Prize and may augment the text.

Nobel Foundation Directory.
> A booklet published by the Nobel Foundation containing general information related to the Nobel Prize, descriptions of the Nobel Foundation and the prize-awarding institutions, a list of publications, and a listing of all laureates in all categories.

Nobel: The Man & His Prizes. Third Edition. Edited by the Nobel Foundation and W. Odelberg. American Elsevier, New York, 1972.
> Contains writings about Alfred Nobel's life and the story of the Nobel Prize, with systematic discussions of the prizes in the original five fields. It also contains the full text of the *Statutes.*

BLOKH, A. M., *The Soviet Union in the Interior of the Nobel Prizes: Facts, Documents, Considerations, Commentaries* (in Russian). Humanistica, St. Peterburg, 2001.
> This is a meticulously documented history of Soviet dealings in connection with the Nobel Prize based on archival material of Soviet governmental and party institutions and other Russian and Swedish documents.

CRAWFORD, E., *The Beginning of the Nobel Institution: The Science Prizes, 1901–1915.* Cambridge University Press, and Editions de la Maison des Sciences de l'Homme, Paris, 1984.
> It provides a detailed history of the first 15 years of the physics and chemistry prizes.

CRAWFORD, E., *Nationalism and Internationalism in Science, 1880–1939: Four Studies of the Nobel Population.* Cambridge University Press, 1992.
> The author examines the internationalization of science and the role of the Nobel Prize in it.

CRAWFORD, E.; HEILBRON, J. L.; ULRICH, R., *The Nobel Population 1901–1937: A Census of the Nominators and Nominees for the Prizes in Physics and Chemistry.* Office for History of Science and Technology, University of California, Berkeley and Office for History of Science, Uppsala University, 1987.
> This is a meticulous and instructive compilation of who nominated whom and vice versa.

FELDMAN, B., *The Nobel Prize: A History of Genius, Controversy, and Prestige*. Arcade Publishing, New York, 2000.

>A survey of all the Nobel Prizes and the Nobel Memorial Prize over the first hundred years, for the general reader.

FRIEDMAN, R. M., *The Politics of Excellence: Behind the Nobel Prize in Science*. Times Books, New York, 2001.

>It focuses on the physics and chemistry prizes during the first 50 years of the Nobel institution and uses extensively archival material of the Royal Swedish Academy of Sciences.

WILHELM, P., *The Nobel Prize*. Teknowledge, Stockholm, 1983.

>This is largely a picture book, supported by the Nobel Foundation, about Alfred Nobel and the Nobel Foundation, using selected examples of laureates and the celebrations.

ZUCKERMAN, H., *Scientific Elite: Nobel Laureates in the United States*. Transactions Publishers, New Brunswick, New Jersey, 1996 (originally by The Free Press, 1977).

>The definitive work of the sociology of science based on the careers of all American Nobel laureates between 1907 and 1972.

Nobel laureates in the sciences, 1901–2002

(Based on the *Nobel Foundation Directory*)

Physics

1901
Wilhelm Conrad Röntgen (1845–1923) 'in recognition of the extraordinary services he has rendered by the discovery of the remarkable rays subsequently named after him'.

1902
Hendrık Antoon Lorentz (1853–1928) and Pieter Zeeman (1865–1943) 'in recognition of the extraordinary service they rendered by their researches into the influence of magnetism upon radiation phenomena'.

1903
Antoine Henri Becquerel (1852–1908) 'in recognition of the extraordinary services he has rendered by his discovery of spontaneous radioactivity'.

Pierre Curie (1859–1906) and Marie Curie, née Sklodowska (1867–1934) 'in recognition of the extraordinary services they have rendered by their joint researches on the radiation phenomena discovered by Professor Henri Becquerel'.

1904
Lord Rayleigh (John William Strutt) (1842–1919) 'for his investigations of the densities of the most important gases and for his discovery of argon in connection with these studies'.

1905
Philipp Eduard Anton Lenard (1862–1947) 'for his work on cathode rays'.

1906
Joseph John Thomson (1856–1940) 'in recognition of the great merits of his theoretical and experimental investigations on the conduction of electricity by gases'.

1907
Albert Abraham Michelson (1852–1931) 'for his optical precision instruments and the spectroscopic and metrological investigations carried out with their aid'.

1908
Gabriel Lippmann (1845 1921) 'for his method of reproducing colours photographically based on the phenomenon of interference'.

1909
Guglielmo Marconi (1874–1937) and Carl Ferdinand Braun (1850–1918) 'in recognition of their contributions to the development of wireless telegraphy'.

1910
Johannes Diderik van der Waals (1837–1923) 'for his work on the equation of state for gases and liquids'.

1911
Wilhelm Wien (1864–1928) 'for his discoveries regarding the laws governing the radiation of heat'.

1912
Nils Gustaf Dalén (1869–1937) 'for his invention of automatic regulators for use in conjunction with gas accumulators for illuminating lighthouses and buoys'.

1913
Heike Kamerlingh-Onnes (1853–1926) 'for his investigations on the properties of matter at low temperatures which led, inter alia, to the production of liquid helium'.

1914
Max von Laue (1879–1960) 'for his discovery of the diffraction of X-rays by crystals'.

1915
William Henry Bragg (1862–1942) and William Lawrence Bragg (1890–1971) 'for their services in the analysis of crystal structure by means of X-rays'.

1918
The prize for 1917:
Charles Glover Barkla (1877–1944) 'for his discovery of the characteristic Röntgen radiation of the elements'.

1919
The prize for 1918:
Max Karl Ernst Ludwig Planck (1858–1947) 'in recognition of the services he rendered to the advancement of Physics by his discovery of energy quanta'.

The prize for 1919:
Johannes Stark (1874–1957) 'for his discovery of the Doppler effect in canal rays and the splitting of spectral lines in electric fields'.

1920
Charles Edouard Guillaume (1861–1938) 'in recognition of the service he has rendered to precision measurements in Physics by his discovery of anomalies in nickel steel alloys'.

1922
The prize for 1921:
Albert Einstein (1879–1955) 'for his services to Theoretical Physics, and especially for his discovery of the law of the photoelectric effect'.

The prize for 1922:
Niels Bohr (1885–1962) 'for his services in the investigation of the structure of atoms and of the radiation emanating from them'.

1923
Robert Andrews Millikan (1868–1953) 'for his work on the elementary charge of electricity and on the photoelectric effect'.

1925
The prize for 1924:
Karl Manne Georg Siegbahn (1886–1978) 'for his discoveries and research in the field of X-ray spectroscopy'.

1926
The prize for 1925:
James Franck (1882–1964) and Gustav Hertz (1887–1975) 'for their discovery of the laws governing the impact of an electron upon an atom'.

The prize for 1926:
Jean Baptiste Perrin (1870–1942) 'for his work on the discontinuous structure of matter, and especially for his discovery of sedimentation equilibrium'.

1927
Arthur Holly Compton (1892–1962) 'for his discovery of the effect named after him'.

Charles Thomson Rees Wilson (1869–1959) 'for his method of making the paths of electrically charged particles visible by condensation of vapour'.

1929
The prize for 1928:
Owen Willans Richardson (1879–1959) 'for his work on the thermionic phenomenon and especially for the discovery of the law named after him'.

The prize for 1929:
Prince Louis-Victor de Broglie (1892–1987) 'for his discovery of the wave nature of electrons'.

1930
Chandrasekhara Venkata Raman (1888–1970) 'for his work on the scattering of light and for the discovery of the effect named after him'.

1933
The prize for 1932:
Werner Heisenberg (1901–76) 'for the creation of quantum mechanics, the application of which has, inter alia, led to the discovery of the allotropic forms of hydrogen'.

The prize for 1933:
Erwin Schrödinger (1887–1961) and Paul Adrien Maurice Dirac (1902–84) 'for the discovery of new productive forms of atomic theory'.

1935
James Chadwick (1891–1974) 'for the discovery of the neutron'.

1936
Victor Franz Hess (1883–1964) 'for his discovery of cosmic radiation'.

Carl David Anderson (1905–91) 'for his discovery of the positron'.

1937
Clinton Joseph Davisson (1881–1958) and George Paget Thomson (1892–1975) 'for their experimental discovery of the diffraction of electrons by crystals'.

1938
Enrico Fermi (1901–54) 'for his demonstrations of the existence of new radioactive elements produced by neutron irradiation, and for his related discovery of nuclear reactions brought about by slow neutrons'.

1939
Ernest Orlando Lawrence (1901–58) 'for the invention and development of the cyclotron and for results obtained with it, especially with regard to artificial radioactive elements'.

1944
The prize for 1943:
Otto Stern (1888–1969) 'for his contribution to the development of the molecular ray method and his discovery of the magnetic moment of the proton'.

The prize for 1944:
Isidor Isaac Rabi (1898–1988) 'for his resonance method for recording the magnetic properties of atomic nuclei'.

1945
Wolfgang Pauli (1900–58) 'for the discovery of the Exclusion Principle, also called the Pauli Principle'.

1946
Percy Williams Bridgman (1882–1961) 'for the invention of an apparatus to produce extremely high pressures, and for the discoveries he made therewith in the field of high pressure physics'.

1947
Edward Victor Appleton (1892–1965) 'for his investigations of the physics of the upper atmosphere especially for the discovery of the so-called Appleton layer'.

1948
Patrick Maynard Stuart Blackett (1897–1974) 'for his development of the Wilson cloud chamber method, and his discoveries therewith in the fields of nuclear physics and cosmic radiation'.

1949
Hideki Yukawa (1907–81) 'for his prediction of the existence of mesons on the basis of theoretical work on nuclear forces'.

1950
Cecil Frank Powell (1903–69) 'for his development of the photographic method of studying nuclear processes and his discoveries regarding mesons made with this method'.

1951
John Douglas Cockcroft (1897–1967) and Ernest Thomas Sinton Walton (1903–95) 'for their pioneer work on the transmutation of atomic nuclei by artificially accelerated atomic particles'.

1952
Felix Bloch (1905–83) and Edward Mills Purcell (1912–97) 'for their development of new methods for nuclear magnetic precision measurements and discoveries in connection therewith'.

1953
Frits (Frederik) Zernike (1888–1966) 'for his demonstration of the phase contrast method, especially for his invention of the phase contrast microscope'.

1954
Max Born (1882–1970) 'for his fundamental research in quantum mechanics, especially for his statistical interpretation of the wavefunction'.

Walther Bothe (1891–1957) 'for the coincidence method and his discoveries made therewith'.

1955
Willis Eugene Lamb (1913–) 'for his discoveries concerning the fine structure of the hydrogen spectrum'.

Polykarp Kusch (1911–93) 'for his precision determination of the magnetic moment of the electron'.

1956
William Shockley (1910–89), John Bardeen (1908–91), and Walter Houser Brattain (1902–87) 'for their researches on semiconductors and their discovery of the transistor effect'.

1957
Chen Ning Yang (1922–) and Tsung-Dao Lee (1926–) 'for their penetrating investigation of the so-called parity laws which has led to important discoveries regarding the elementary particles'.

1958
Pavel Alekseyevich Cherenkov (1904–90), Il'ja Mikhailovich Frank (1908–90), and Igor Yevgenyevich Tamm (1885–1971) 'for the discovery and the interpretation of the Cherenkov effect'.

1959
Emilio Gino Segrè (1905–89) and Owen Chamberlain (1920–) 'for their discovery of the antiproton'.

1960
Donald A. Glaser (1926–) 'for the invention of the bubble chamber'.

1961
Robert Hofstadter (1915–90) 'for his pioneering studies of electron scattering in atomic nuclei and for his thereby achieved discoveries concerning the structure of the nucleons'.

Rudolf Ludwig Mössbauer (1929–) 'for his researches concerning the resonance absorption of gamma radiation and his discovery in this connection of the effect which bears his name'.

1962
Lev Davidovich Landau (1908–68) 'for his pioneering theories for condensed matter, especially liquid helium'.

1963
Eugene P. Wigner (1902–95) 'for his contributions to the theory of the atomic nucleus and the elementary particles, particularly through the discovery and application of fundamental symmetry principles'.

Maria Goeppert-Mayer (1906–72) and J. Hans D. Jensen (1907–73) 'for their discoveries concerning nuclear shell structure'.

1964
Charles H. Townes (1915–)

Nicolay Gennadiyevich Basov (1922–2001) and Aleksandr Mikhailovich Prokhorov (1916–2002)
'for fundamental work in the field of quantum electronics, which has led to the construction of oscillators and amplifiers based on the maser–laser principle'.

1965
Sin-Itiro Tomonaga (1906–79), Julian Schwinger (1918–94), and Richard P. Feynman (1918–88) 'for their fundamental work in quantum electrodynamics, with deep-ploughing consequences for the physics of elementary particles'.

1966
Alfred Kastler (1902–84) 'for the discovery and development of optical methods for studying hertzian resonances in atoms'.

1967
Hans Albrecht Bethe (1906–) 'for his contributions to the theory of nuclear reactions, especially his discoveries concerning the energy production in stars'.

1968
Luis W. Alvarez (1911–88) 'for his decisive contributions to elementary particle physics, in particular the discovery of a large number of resonance states, made possible through his development of the technique of using hydrogen bubble chamber and data analysis'.

1969
Murray Gell-Mann (1929–) 'for his contributions and discoveries concerning the classification of elementary particles and their interactions'.

1970
Hannes Alfvén (1908–95) 'for fundamental work and discoveries in magnetohydrodynamics with fruitful applications in different parts of plasma physics'.

Louis Néel (1904–2000) 'for fundamental work and discoveries concerning antiferromagnetism and ferrimagnetism which have led to important applications in solid state physics'.

1971
Dennis Gabor (1900–79) 'for his invention and development of the holographic method'.

1972
John Bardeen (1908–91), Leon N. Cooper (1930–), and J. Robert Schrieffer (1931–) 'for their jointly developed theory of superconductivity, usually called the BCS-theory'.

1973
Leo Esaki (1925–) and Ivar Giaever (1929–) 'for their experimental discoveries regarding tunneling phenomena in semiconductors and superconductors, respectively'.

Brian D. Josephson (1940–) 'for his theoretical predictions of the properties of a supercurrent through a tunnel barrier, in particular those phenomena which are generally known as the Josephson effects'.

1974
Martin Ryle (1918–84) and Antony Hewish (1924–) 'for their pioneering research in radio astrophysics: Ryle for his observations and inventions, in particular of the aperture synthesis technique, and Hewish for his decisive role in the discovery of pulsars'.

1975
Aage Bohr (1922–), Ben Mottelson (1926–), and James Rainwater (1917–86) 'for the discovery of the connection between collective motion and particle motion in atomic nuclei and the development of the theory of the structure of the atomic nucleus based on this connection'.

1976
Burton Richter (1931–) and Samuel C. C. Ting (1936–) 'for their pioneering work in the discovery of a heavy elementary particle of a new kind'.

1977
Philip W. Anderson (1923–), Nevill F. Mott (1905–96), and John H. Van Vleck (1899–1980) 'for their fundamental theoretical investigations of the electronic structure of magnetic and disordered systems'.

1978
Pyotr Leonidovich Kapitsa (1894–1984) 'for his basic inventions and discoveries in the area of low-temperature physics'.

Arno A. Penzias (1933–) and Robert W. Wilson (1936–) 'for their discovery of cosmic microwave background radiation'.

1979
Sheldon L. Glashow (1932–), Abdus Salam (1926–96), and Steven Weinberg (1933–) 'for their contributions to the theory of the unified weak and electromagnetic interaction between elementary particles, including, inter alia, the prediction of the weak neutral current'.

1980
James W. Cronin (1931–) and Val L. Fitch (1923–) 'for the discovery of violations of fundamental symmetry principles in the decay of neutral K-mesons'.

1981
Nicolaas Bloembergen (1920–) and Arthur L. Schawlow (1921–99) 'for their contribution to the development of laser spectroscopy'.

Kai M. Siegbahn (1918–) 'for his contribution to the development of high-resolution electron spectroscopy'.

1982
Kenneth G. Wilson (1936–) 'for his theory for critical phenomena in connection with phase transitions'.

1983
Subramanyan Chandrasekhar (1910–95) 'for his theoretical studies of the physical processes of importance to the structure and evolution of the stars'.

William A. Fowler (1911–95) 'for his theoretical and experimental studies of the nuclear reactions of importance in the formation of the chemical elements in the universe'.

1984
Carlo Rubbia (1934–) and Simon van der Meer (1925–) 'for their decisive contributions to the large project, which led to the discovery of the field particles W and Z, communicators of weak interaction'.

1985
Klaus von Klitzing (1943–) 'for the discovery of the quantized Hall effect'.

1986
Ernst Ruska (1906–88) 'for his fundamental work in electron optics, and for the design of the first electron microscope'.

Gerd Binnig (1947–) and Heinrich Rohrer (1933–) 'for their design of the scanning tunneling microscope'.

1987
J. Georg Bednorz (1950–) and K. Alexander Müller (1927–) 'for their important break-through in the discovery of superconductivity in ceramic materials'.

1988
Leon M. Lederman (1922–), Melvin Schwartz (1932–), and Jack Steinberger (1921–) 'for the neutrino beam method and the demonstration of the doublet structure of the leptons through the discovery of the muon neutrino'.

1989

Norman F. Ramsey (1915–) 'for the invention of the separated oscillatory fields method and its use in the hydrogen maser and other atomic clocks'.

Hans G. Dehmelt (1922–) and Wolfgang Paul (1913–93) 'for the development of the ion trap technique'.

1990

Jerome I. Friedman (1930–), Henry W. Kendall (1926–1999), and Richard E. Taylor (1929–) 'for their pioneering investigations concerning deep inelastic scattering of electrons on protons and bound neutrons, which have been of essential importance for the development of the quark model in particle physics'.

1991

Pierre-Gilles de Gennes (1932–) 'for discovering that methods developed for studying order phenomena in simple systems can be generalized to more complex forms of matter, in particular to liquid crystals and polymers'.

1992

Georges Charpak (1924–) 'for his invention and development of particle detectors, in particular the multiwire proportional chamber'.

1993

Russell A. Hulse (1950–) and Joseph H. Taylor Jr. (1941–) 'for the discovery of a new type of pulsar, a discovery that has opened up new possibilities for the study of gravitation'.

1994

'for pioneering contributions to the development of neutron scattering techniques for studies of condensed matters'.

Bertram N. Brockhouse (1918–) 'for the development of neutron spectroscopy'.

Clifford G. Shull (1915–2001) 'for the development of the neutron diffraction technique'.

1995

'for pioneering experimental contributions to lepton physics'.

Martin L. Perl (1927–) 'for the discovery of the tau lepton'.

Frederick Reines (1918–1998) 'for the detection of the neutrino'.

1996

David M. Lee (1931–), Douglas D. Osheroff (1945–), and Robert C. Richardson (1937–) 'for their discovery of superfluidity in helium-3'.

1997
Steven Chu (1948–), Claude Cohen-Tannoudji (1933–), and William D. Phillips (1948–) 'for development of methods to cool and trap atoms with laser light'.

1998
Robert B. Laughlin (1950–), Horst L. Störmer (1949–), and Daniel C. Tsui (1939–) 'for their discovery of a new form of quantum fluid with fractionally charged excitation'.

1999
Gerardus 't Hooft (1946) and Martinus J. G. Veltman (1931) 'for elucidating the quantum structure of electroweak interactions in physics'.

2000
'for basic work on information and communication technology'.

Jack S. Kilby (1923–) 'for his part in the invention of the integrated circuit'.

Zhores I. Alferov (1930–) and Herbert Kroemer (1928–) 'for developing semiconductor heterostructures used in high-speed- and opto-electronics'.

2001
Eric A. Cornell (1961–), Wolfgang Ketterle (1957–), and Carl E. Wieman (1951–) 'for the achievement of Bose-Einstein condensation in dilute gases of alkali atoms, and for early fundamental studies of the properties of the condensates'.

2002
Raymond Davis, Jr. (1914–) and Masatoshi Koshiba (1926–) 'for pioneering contributions to astrophysics, in particular for the detection of cosmic neutrinos'.

Riccardo Giacconi (1931–) 'for pioneering contributions to astrophysics, which have led to the discovery of cosmic X-ray sources'.

Chemistry

1901
Jacobus Henricus van 't Hoff (1852–1911) 'in recognition of the extra-ordinary services he has rendered by the discovery of the laws of chemical dynamics and osmotic pressure in solutions'.

1902
Hermann Emil Fischer (1852–1919) 'in recognition of the extraordinary services he has rendered by his work on sugar and purine syntheses'.

1903
Svante August Arrhenius (1859–1927) 'in recognition of the extraordinary services he has rendered to the advancement of chemistry by his electrolytic theory of dissociation'.

1904
William Ramsay (1852–1916) 'in recognition of his services in the discovery of the inert gaseous elements in air, and his determination of their place in the periodic system'.

1905
Johann Friedrich Wilhelm Adolf von Baeyer (1835–1917) 'in recognition of his services in the advancement of organic chemistry and the chemical industry, through his work on organic dyes and hydroaromatic compounds'.

1906
Henri Moissan (1852–1907) 'in recognition of the great services rendered by him in his investigation and isolation of the element fluorine, and for the adoption in the service of science of the electric furnace called after him'.

1907
Eduard Buchner (1860–1917) 'for his biochemical researches and his discovery of cell-free fermentation'.

1908
Ernest Rutherford (1871–1937) 'for his investigations into the disintegration of the elements, and the chemistry of radioactive substances'.

1909
Wilhelm Ostwald (1853–1932) 'in recognition of his work on catalysis and for his investigations into the fundamental principles governing chemical equilibria and rates of reaction'.

1910
Otto Wallach (1847–1931) 'in recognition of his services to organic chemistry and the chemical industry by his pioneer work in the field of alicyclic compounds'.

1911
Marie Curie, née Sklodowska (1867–1934) 'in recognition of her services to the advancement of chemistry by the discovery of the elements radium and polonium, by the isolation of radium and the study of the nature and compounds of this remarkable element'.

1912
Victor Grignard (1871–1935) 'for the discovery of the so-called Grignard reagent, which in recent years has greatly advanced the progress of organic chemistry'.

Paul Sabatier (1854–1941) 'for his method of hydrogenating organic compounds in the presence of finely disintegrated metals whereby the progress of organic chemistry has been greatly advanced in recent years'.

1913
Alfred Werner (1866–1919) 'in recognition of his work on the linkage of atoms in molecules by which he has thrown new light on earlier investigations and opened up new fields of research especially in inorganic chemistry'.

1915
The prize for 1914:
Theodore William Richards (1868–1928) 'in recognition of his accurate determinations of the atomic weight of a large number of chemical elements'.

The prize for 1915:
Richard Martin Willstätter (1872–1942) 'for his researches on plant pigments, especially chlorophyll'.

1919
The prize for 1918:
Fritz Haber (1868–1934) 'for the synthesis of ammonia from its elements'.

1921
The prize for 1920:
Walther Hermann Nernst (1864–1941) 'in recognition of his work in thermochemistry'.

1922
The prize for 1921:
Frederick Soddy (1877–1956) 'for his contributions to our knowledge of the chemistry of radioactive substances, and his investigations into the origin and nature of isotopes'.

The prize for 1922:
Francis William Aston (1877–1945) 'for his discovery, by means of his mass spectrograph, of isotopes, in a large number of non-radioactive elements, and for his enunciation of the whole-number rule'.

1923
Fritz Pregl (1869–1930) 'for his invention of the method of micro-analysis of organic substances'.

1926
The prize for 1925:
Richard Adolf Zsigmondy (1865–1929) 'for his demonstration of the heterogeneous nature of colloid solutions and for the methods he used, which have since become fundamental in modern colloid chemistry'.

The prize for 1926:
The (Theodor) Svedberg (1884–1971) 'for his work on disperse systems'.

1928
The prize for 1927:
Heinrich Otto Wieland (1877–1957) 'for his investigations of the constitution of the bile acids and related substances'.

The prize for 1928:
Adolf Otto Reinhold Windaus (1876–1959) 'for the services rendered through his research into the constitution of the sterols and their connection with the vitamins'.

1929
Arthur Harden (1865–1940) and Hans Karl August Simon von Euler-Chelpin (1873–1964) 'for their investigations on the fermentation of sugar and fermentative enzymes'.

1930
Hans Fischer (1881–1945) 'for his researches into the constitution of haemin and chlorophyll and especially for his synthesis of haemin'.

1931
Carl Bosch (1874–1940) and Friedrich Bergius (1884–1949) 'in recognition of their contributions to the invention and development of chemical high pressure methods'.

1932
Irving Langmuir (1881–1957) 'for his discoveries and investigations in surface chemistry'.

1934
Harold Clayton Urey (1893–1981) 'for his discovery of heavy hydrogen'.

1935
Frédéric Joliot (1900–58) and Irène Joliot-Curie (1897–1956) 'in recognition of their synthesis of new radioactive elements'.

1936
Petrus (Peter) Josephus Wilhelmus Debye (1884–1966) 'for his contributions to our knowledge of molecular structure through his investigations on dipole moments and on the diffraction of X-rays and electrons in gases'.

1937
Walter Norman Haworth (1883–1950) 'for his investigations on carbohydrates and vitamin C'.

Paul Karrer (1889–1971) 'for his investigations on carotenoids, flavins and vitamins A and B2'.

1939
The prize for 1938:
Richard Kuhn (1900–67) 'for his work on carotenoids and vitamins'.

The prize for 1939:
Adolf Friedrich Johann Butenandt (1903–95) 'for his work on sex hormones'.

Leopold Ružička (1887–1976) 'for his work on polymethylenes and higher terpenes'.

1944
The prize for 1943:
George de Hevesy (1885–1966) 'for his work on the use of isotopes as tracers in the study of chemical processes'.

1945
The prize for 1944:
Otto Hahn (1879–1968) 'for his discovery of the fission of heavy nuclei'.

The prize for 1945:
Artturi Ilmari Virtanen (1895–1973) 'for his research and inventions in agricultural and nutrition chemistry, especially for his fodder preservation method'.

1946
James Batcheller Sumner (1887–1955) 'for his discovery that enzymes can be crystallized'.

John Howard Northrop (1891–1987) and Wendell Meredith Stanley (1904–71) 'for their preparation of enzymes and virus proteins in a pure form'.

1947
Robert Robinson (1886–1975) 'for his investigations on plant products of biological importance, especially the alkaloids'.

1948
Arne Wilhelm Kaurin Tiselius (1902–71) 'for his research on electrophoresis and adsorption analysis, especially for his discoveries concerning the complex nature of serum proteins'.

1949
William Francis Giauque (1895–1982) 'for his contributions in the field of chemical thermodynamics, particularly concerning the behaviour of substances at extremely low temperatures'.

1950
Otto Paul Hermann Diels (1876–1954) and Kurt Alder (1902–58) 'for their discovery and development of the diene synthesis'.

1951
Edwin Mattison McMillan (1907–91) and Glenn Theodore Seaborg (1912–99) 'for their discoveries in the chemistry of the transuranium elements'.

1952
Archer John Porter Martin (1910–2002) and Richard Laurence Millington Synge (1914–94) 'for their invention of partition chromatography'.

1953
Hermann Staudinger (1881–1965) 'for his discoveries in the field of macromolecular chemistry'.

1954
Linus Carl Pauling (1901–94) 'for his research into the nature of the chemical bond and its application to the elucidation of the structure of complex substances'.

1955
Vincent du Vigneaud (1901–78) 'for his work on biochemically important sulphur compounds, especially for the first synthesis of a polypeptide hormone'.

1956
Cyril Norman Hinshelwood (1897–1967) and Nikolay Nikolaevich Semenov (1896–1986) 'for their researches into the mechanism of chemical reactions'.

1957
Alexander R. Todd (1907–97) 'for his work on nucleotides and nucleotide coenzymes'.

1958
Frederick Sanger (1918–) 'for his work on the structure of proteins, especially that of insulin'.

1959
Jaroslav Heyrovsky (1890–1967) 'for his discovery and development of the polarographic methods of analysis'.

1960
Willard Frank Libby (1908–80) 'for his method to use carbon-14 for age determination in archaeology, geology, geophysics, and other branches of science'.

1961
Melvin Calvin (1911–97) 'for his research on the carbon dioxide assimilation in plants'.

1962
Max Ferdinand Perutz (1914–2002) and John Cowdery Kendrew (1917–97) 'for their studies of the structures of globular proteins'.

1963
Karl Ziegler (1898–1973) and Giulio Natta (1903–79) 'for their discoveries in the field of the chemistry and technology of high polymers'.

1964
Dorothy Crowfoot Hodgkin (1910–94) 'for her determinations by X-ray techniques of the structures of important biochemical substances'.

1965
Robert Burns Woodward (1917–79) 'for his outstanding achievements in the art of organic synthesis'.

1966
Robert S. Mulliken (1896–1986) 'for his fundamental work concerning chemical bonds and the electronic structure of molecules by the molecular orbital method'.

1967
Manfred Eigen (1927–)

Ronald George Wreyford Norrish (1897–1978) and George Porter (1920–2002) 'for their studies of extremely fast chemical reactions, effected by disturbing the equilibrium by means of very short pulses of energy'.

1968
Lars Onsager (1903–76) 'for the discovery of the reciprocal relations bearing his name, which are fundamental for the thermodynamics of irreversible processes'.

1969
Derek H. R. Barton (1918–98) and Odd Hassel (1897–1981) 'for their contributions to the development of the concept of conformation and its application in chemistry'.

1970
Luis F. Leloir (1906–87) 'for his discovery of sugar nucleotides and their role in the biosynthesis of carbohydrates'.

1971
Gerhard Herzberg (1904–99) 'for his contributions to the knowledge of electronic structure and geometry of molecules, particularly free radicals'.

1972
Christian B. Anfinsen (1916–95) 'for his work on ribonuclease, especially concerning the connection between the amino acid sequence and the biologically active conformation'.

Stanford Moore (1913–82) and William H. Stein (1911–80) 'for their contribution to the understanding of the connection between chemical structure and catalytic activity of the active centre of the ribonuclease molecule'.

1973
Ernst Otto Fischer (1918–) and Geoffrey Wilkinson (1921–96) 'for their pioneering work, performed independently, on the chemistry of the organometallic, so called sandwich compounds'.

1974
Paul J. Flory (1910–85) 'for his fundamental achievements, both theoretical and experimental, in the physical chemistry of the macromolecules'.

1975
John Warcup Cornforth (1917–) 'for his work on the stereochemistry of enzyme-catalyzed reactions'.

Vladimir Prelog (1906–98) 'for his research into the stereochemistry of organic molecules and reactions'.

1976
William N. Lipscomb (1919–) 'for his studies on the structure of boranes illuminating problems of chemical bonding'.

1977
Ilya Prigogine (1917–) 'for his contributions to non-equilibrium thermodynamics, particularly the theory of dissipative structures'.

1978
Peter D. Mitchell (1920–92) 'for his contribution to the understanding of biological energy transfer through the formulation of the chemiosmotic theory'.

1979
Herbert C. Brown (1912–) and Georg Wittig (1897–1987) 'for their development of the use of boron- and phosphorus-containing compounds, respectively, into important reagents in organic synthesis'.

1980
Paul Berg (1926–) 'for his fundamental studies of the biochemistry of nucleic acids, with particular regard to recombinant-DNA'.

Walter Gilbert (1932–) and Frederick Sanger (1918–) 'for their contributions concerning the determination of base sequences in nucleic acids'.

1981
Kenichi Fukui (1918–98) and Roald Hoffmann (1937–) 'for their theories, developed independently, concerning the course of chemical reactions'.

1982
Aaron Klug (1926–) 'for his development of crystallographic electron microscopy and his structural elucidation of biologically important nucleic acid–protein complexes'.

1983
Henry Taube (1915–) 'for his work on the mechanisms of electron transfer reactions, especially in metal complexes'.

1984
Robert Bruce Merrifield (1921–) 'for his development of methodology for chemical synthesis on a solid matrix'.

1985
Herbert A. Hauptman (1917–) and Jerome Karle (1918–) 'for their outstanding achievements in the development of direct methods for the determination of crystal structures'.

1986
Dudley R. Herschbach (1932–), Yuan T. Lee (1936–), and John C. Polanyi (1929–) 'for their contributions concerning the dynamics of chemical elementary processes'.

1987
Donald J. Cram (1919–2001), Jean-Marie Lehn (1939–), and Charles J. Pedersen (1904–89) 'for their development and use of molecules with structure-specific interactions of high selectivity'.

1988
Johann Deisenhofer (1943–), Robert Huber (1937–), and Hartmut Michel (1948–) 'for the determination of the three-dimensional structure of a photosynthetic reaction centre'.

1989
Sidney Altman (1939–) and Thomas R. Cech (1947–) 'for their discovery of catalytic properties of RNA'.

1990
Elias James Corey (1928–) 'for his development of the theory and methodology of organic synthesis'.

1991
Richard R. Ernst (1933–) 'for his contributions to the development of the methodology of high resolution nuclear magnetic resonance (NMR) spectroscopy'.

1992
Rudolph A. Marcus (1923–) 'for his contributions to the theory of electron transfer reactions in chemical systems'.

1993
'for contributions to the developments of methods within DNA-based chemistry'.

Kary B. Mullis (1944–) 'for his invention of the polymerase chain reaction (PCR) method'.

Michael Smith (1932–2000) 'for his fundamental contributions to the establishment of oligonucleotide-based, site-directed mutagenesis and its development for protein studies'.

1994
George A. Olah (1927–) 'for his contribution to carbocation chemistry'.

1995
Paul J. Crutzen (1933–), Mario J. Molina (1943–), and F. Sherwood Rowland (1927–) 'for their work in atmospheric chemistry, particularly concerning the formation and decomposition of ozone'.

1996
Robert F. Curl Jr. (1933–), Harold W. Kroto (1939–), and Richard E. Smalley (1943–) 'for their discovery of fullerenes'.

1997
Paul D. Boyer (1918–) and John E. Walker (1941–) 'for their elucidation of the enzymatic mechanism underlying the synthesis of adenosine triphosphate (ATP)'.

Jens C. Skou (1918–) 'for the first discovery of an ion-transporting enzyme, Na^+, K^+-ATPase'.

1998
Walter Kohn (1923–) 'for his development of the density-functional theory'.

John A. Pople (1925–) for his development of computational methods in quantum chemistry'.

1999
Ahmed Zewail (1946–) 'for his studies of the transition states of chemical reactions using femtosecond spectroscopy'.

2000
Alan J. Heeger (1936–), Alan G. MacDiarmid (1927–), and Hideki Shirakawa (1936–) 'for the discovery and development of conductive polymers'.

2001
William S. Knowles (1917–) and Ryoji Noyori (1938–) 'for their work on chirally catalysed hydrogenation reactions'.

K. Barry Sharpless (1941–) 'for his work on chirally catalysed oxidation reactions'.

2002

'for the development of methods for identification and structure analyses of biological macromolecules'.

John B. Fenn (1917–) and Koichi Tanaka (1959–) 'for their development of soft desorption ionization methods for mass spectrometric analyses of biological macromolecules'.

Kurt Wüthrich (1938–) 'for his development of nuclear magnetic resonance spectroscopy for determining the three-dimensional structure of biological macromolecules in solution'.

Physiology or medicine

1901

Emil Adolf von Behring (1854–1917) 'for his work on serum therapy, especially its application against diphtheria, by which he has opened a new road in the domain of medical science and thereby placed in the hands of the physician a victorious weapon against illness and deaths'.

1902

Ronald Ross (1857–1932) 'for his work on malaria, by which he has shown how it enters the organism and thereby has laid the foundation for successful research on this disease and methods of combating it'.

1903

Niels Ryberg Finsen (1860–1904) 'in recognition of his contribution to the treatment of diseases, especially lupus vulgaris, with concentrated light radiation, whereby he has opened a new avenue for medical science'.

1904

Ivan Petrovich Pavlov (1849–1936) 'in recognition of his work on the physiology of digestion, through which knowledge on vital aspects of the subject has been transformed and enlarged'.

1905

Robert Koch (1843–1910) 'for his investigations and discoveries in relation to tuberculosis'.

1906

Camillo Golgi (1843–1926) and Santiago Ramón y Cajal (1852–1934) 'in recognition of their work on the structure of the nervous system'.

1907

Charles Louis Alphonse Laveran (1845–1922) 'in recognition of his work on the role played by protozoa in causing diseases'.

1908

Ilya Ilyich Mechnikov (1845–1916) and Paul Ehrlich (1854–1915) 'in recognition of their work on immunity'.

1909

Emil Theodor Kocher (1841–1917) 'for his work on the physiology, pathology and surgery of the thyroid gland'.

1910

Albrecht Kossel (1853–1927) 'in recognition of the contributions to our knowledge of cell chemistry made through his work on proteins, including the nucleic substances'.

1911

Allvar Gullstrand (1862–1930) 'for his work on the dioptrics of the eye'.

1912

Alexis Carrel (1873–1944) 'in recognition of his work on vascular suture and the transplantation of blood vessels and organs'.

1913

Charles Robert Richet (1850–1935) 'in recognition of his work on anaphylaxis'.

1914

Robert Bárány (1876–1936) 'for his work on the physiology and pathology of the vestibular apparatus'.

1920

The prize for 1919:
Jules Bordet (1870–1961) 'for his discoveries relating to immunity'.

The prize for 1920:
Schack August Steenberg Krogh (1874–1949) 'for his discovery of the capillary motor regulating mechanism'.

1923

The prize for 1922:
Archibald Vivian Hill (1886–1977) 'for his discovery relating to the production of heat in the muscle'.

Otto Fritz Meyerhof (1884–1951) 'for his discovery of the fixed relationship between the consumption of oxygen and the metabolism of lactic acid in the muscle'.

The prize for 1923:
Frederick Grant Banting (1891–1941) and John James Richard Macleod (1876–1935) 'for the discovery of insulin'.

1924
Willem Einthoven (1860–1927) 'for his discovery of the mechanism of the electrocardiogram'.

1927
The prize for 1926:
Johannes Andreas Grib Fibiger (1867–1928) 'for his discovery of the Spiroptera carcinoma'.

The prize for 1927:
Julius Wagner-Jauregg (1857–1940) 'for his discovery of the therapeutic value of malaria inoculation in the treatment of dementia paralytica'.

1928
Charles Jules Henri Nicolle (1866–1936) 'for his work on typhus'.

1929
Christiaan Eijkman (1858–1930) 'for his discovery of the antineuritic vitamin'.

Frederick Gowland Hopkins (1861–1947) 'for his discovery of the growth-stimulating vitamins'.

1930
Karl Landsteiner (1868–1943) 'for his discovery of human blood groups'.

1931
Otto Heinrich Warburg (1883–1970) 'for his discovery of the nature and mode of action of the respiratory enzyme'.

1932
Charles Scott Sherrington (1857–1952) and Edgar Douglas Adrian (1889–1977) 'for their discoveries regarding the functions of neurons'.

1933
Thomas Hunt Morgan (1866–1945) 'for his discoveries concerning the role played by the chromosome in heredity'.

1934
George Hoyt Whipple (1878–1976), George Richards Minot (1885–1950), and William Parry Murphy (1892–1987) 'for their discoveries concerning liver therapy in cases of anaemia'.

1935
Hans Spemann (1869–1941) 'for his discovery of the organizer effect in embryonic development'.

1936
Henry Hallett Dale (1875–1968) and Otto Loewi (1873–1961) 'for their discoveries relating to chemical transmission of nerve impulses'.

1937

Albert von Szent-Györgyi Nagyrapolt (1893–1986) 'for his discoveries in connection with the biological combustion processes, with special reference to vitamin C and the catalysis of fumaric acid'.

1939

The prize for 1938:
Corneille Jean François Heymans (1892–1968) 'for the discovery of the role played by the sinus and aortic mechanisms in the regulation of respiration'.

The prize for 1939:
Gerhard Domagk (1895–1964) 'for the discovery of the antibacterial effects of prontosil'.

1944

The prize for 1943:
Henrik Carl Peter Dam (1895–1976) 'for his discovery of vitamin K'.

Edward Adelbert Doisy (1893–1986) 'for his discovery of the chemical nature of vitamin K'.

The prize for 1944:
Joseph Erlanger (1874–1965) and Herbert Spencer Gasser (1888–1963) 'for their discoveries relating to the highly differentiated functions of single nerve fibres'.

1945

Alexander Fleming (1881–1955), Ernst Boris Chain (1906–79), and Howard Walter Florey (1898–1968) 'for the discovery of penicillin and its curative effect in various infectious diseases'.

1946

Hermann Joseph Muller (1890–1967) 'for the discovery of the production of mutations by means of X-ray irradiation'.

1947

Carl Ferdinand Cori (1896–1984) and Gerty Theresa Cori, née Radnitz (1896–1957) 'for their discovery of the course of the catalytic conversion of glycogen'.

Bernardo Alberto Houssay (1887–1971) 'for his discovery of the part played by the hormone of the anterior pituitary lobe in the metabolism of sugar'.

1948

Paul Hermann Muller (1899–1965) 'for his discovery of the high efficiency of DDT as a contact poison against several arthropods'.

1949

Walter Rudolf Hess (1881–1973) 'for his discovery of the functional organization of the interbrain as a coordinator of the activities of the internal organs'.

Antonio Caetano de Abreu Freire Egas Moniz (1874–1955) 'for his discovery of the therapeutic value of leucotomy in certain psychoses'.

1950

Edward Calvin Kendall (1886–1972), Tadeus Reichstein (1897–1996), and Philip Showalter Hench (1896–1965) 'for their discoveries relating to the hormones of the adrenal cortex, their structure and biological effects'.

1951

Max Theiler (1899–1972) 'for his discoveries concerning yellow fever and how to combat it'.

1952

Selman Abraham Waksman (1888–1973) 'for his discovery of streptomycin, the first antibiotic effective against tuberculosis'.

1953

Hans Adolf Krebs (1900–81) 'for his discovery of the citric acid cycle'.

Fritz Albert Lipmann (1899–1986) 'for his discovery of co-enzyme A and its importance for intermediary metabolism'.

1954

John Franklin Enders (1897–1985), Thomas Huckle Weller (1915–), and Frederick Chapman Robbins (1916–) 'for their discovery of the ability of poliomyelitis viruses to grow in cultures of various types of tissue'.

1955

Axel Hugo Theodor Theorell (1903–82) 'for his discoveries concerning the nature and mode of action of oxidation enzymes'.

1956

André Frédéric Cournand (1895–1988), Werner Forssmann (1904–79), and Dickinson W. Richards (1895–1973) 'for their discoveries concerning heart catheterization and pathological changes in the circulatory system'.

1957

Daniel Bovet (1907–92) 'for his discoveries relating to synthetic compounds that inhibit the action of certain body substances, and especially their action on the vascular system and the skeletal muscles'.

1958

George Wells Beadle (1903–89) and Edward Lawrie Tatum (1909–75) 'for their discovery that genes act by regulating definite chemical events'.

Joshua Lederberg (1925–) 'for his discoveries concerning genetic recombination and the organization of the genetic material of bacteria'.

1959
Severo Ochoa (1905–93) and Arthur Kornberg (1918–) 'for their discovery of the mechanisms in the biological synthesis of ribonucleic acid and deoxyribonucleic acid'.

1960
Frank MacFarlane Burnet (1899–1985) and Peter Brian Medawar (1915–87) 'for discovery of acquired immunological tolerance'.

1961
Georg von Békésy (1899–1972) 'for his discoveries of the physical mechanism of stimulation within the cochlea'.

1962
Francis Harry Compton Crick (1916–), James Dewey Watson (1928–), and Maurice Hugh Frederick Wilkins (1916–) 'for their discoveries concerning the molecular structure of nuclear acids and its significance for information transfer in living material'.

1963
John Carew Eccles (1914–97), Alan Lloyd Hodgkin (1914–98), and Andrew Fielding Huxley (1917–) 'for their discoveries concerning the ionic mechanisms involved in excitation and inhibition in the peripheral and central portions of the nerve cell membrane'.

1964
Konrad Bloch (1912–2000) and Feodor Lynen (1911–79) 'for their discoveries concerning the mechanism and regulation of the cholesterol and fatty acid metabolism'.

1965
François Jacob (1920–), André Lwoff (1902–94), and Jacques Monod (1910–76) 'for their discoveries concerning genetic control of enzyme and virus synthesis'.

1966
Peyton Rous (1879–1970) 'for his discovery of tumour-inducing viruses'.

Charles Brenton Huggins (1901–98) 'for his discoveries concerning hormonal treatment of prostatic cancer'.

1967
Ragnar Granit (1900–91), Haldan Keffer Hartline (1903–83), and George Wald (1906–97) 'for their discoveries concerning the primary physiological and chemical visual processes in the eye'.

1968
Robert W. Holley (1922–93), Har Gobind Khorana (1922–), and Marshall W. Nirenberg (1927–) 'for their interpretation of the genetic code and its function in protein synthesis'.

1969
Max Delbrück (1906–81), Alfred D. Hershey (1908–97), and Salvador E. Luria (1912–91) 'for their discoveries concerning the replication mechanism and the genetic structure of viruses'.

1970
Bernard Katz (1911–), Ulf von Euler (1905–83), and Julius Axelrod (1912–) 'for their discoveries concerning the humoral transmitters in the nerve terminals and the mechanism for their storage, release and inactivation'.

1971
Earl W. Sutherland Jr. (1915–74) 'for his discoveries concerning the mechanisms of the action of hormones'.

1972
Gerald M. Edelman (1929–) and Rodney R. Porter (1917–85) 'for their discoveries concerning the chemical structure of antibodies'.

1973
Karl von Frisch (1886–1982), Konrad Lorenz (1903–89), and Nikolaas Tinbergen (1907–88) 'for their discoveries concerning organization and elicitation of individual and social behaviour patterns'.

1974
Albert Claude (1899–1983), Christian de Duve (1917–), and George E. Palade (1912–) 'for their discoveries concerning the structural and functional organization of the cell'.

1975
David Baltimore (1938–), Renato Dulbecco (1914–), and Howard Martin Temin (1934–94) 'for their discoveries concerning the interaction between tumour viruses and the genetic material of the cell'.

1976
Baruch S. Blumberg (1925–) and D. Carleton Gajdusek (1923–) 'for their discoveries concerning new mechanisms for the origin and dissemination of infectious diseases'.

1977
Roger Guillemin (1924–) and Andrew V. Schally (1926–) 'for their discoveries concerning the peptide hormone production of the brain'.

Rosalyn Yalow (1921–) 'for the development of radioimmunoassays of peptide hormones'.

1978
Werner Arber (1929–), Daniel Nathans (1928–99), and Hamilton O. Smith (1931–) 'for the discovery of restriction enzymes and their application to problems of molecular genetics'.

1979
Allan M. Cormack (1924–) and Godfrey N. Hounsfield (1919–) 'for the development of computer assisted tomography'.

1980
Baruj Benacerraf (1920–), Jean Dausset (1916–), and George D. Snell (1903–96) 'for their discoveries concerning genetically determined structures on the cell surface that regulate immunological reactions'.

1981
Roger W. Sperry (1913–94) 'for his discoveries concerning the functional specialization of the cerebral hemispheres'.

David H. Hubel (1926–) and Torsten N. Wiesel (1924–) 'for their discoveries concerning information processing in the visual system'.

1982
Sune K. Bergström (1916–), Bengt I. Samuelsson (1934–), and John R. Vane (1927–) 'for their discoveries concerning prostaglandins and related biologically active substances'.

1983
Barbara McClintock (1902–92) 'for her discovery of mobile genetic elements'.

1984
Niels K. Jerne (1911–94), Georges J. F. Köhler (1946–95), and César Milstein (1927–2002) 'for theories concerning the specificity in development and control of the immune system and the discovery of the principle for production of monoclonal antibodies'.

1985
Michael S. Brown (1941–) and Joseph L. Goldstein (1940–) 'for their discoveries concerning the regulation of cholesterol metabolism'.

1986
Stanley Cohen (1922–) and Rita Levi-Montalcini (1909–) 'for their discoveries of growth factors'.

1987
Susumu Tonegawa (1939–) 'for his discovery of the genetic principle for generation of antibody diversity'.

1988
James W. Black (1924–), Gertrude B. Elion (1918–99), and George H. Hitchings (1905–98) 'for their discoveries of important principles for drug treatment'.

1989
J. Michael Bishop (1936–) and Harold E. Varmus (1939–) 'for their discovery of the cellular origin of retroviral oncogenes'.

1990
Joseph E. Murray (1919–) and E. Donnall Thomas (1920–) 'for their discoveries concerning organ and cell transplantation in the treatment of human disease'.

1991
Erwin Neher (1944–) and Bert Sakmann (1942–) 'for their discoveries concerning the function of single ion channels in cells'.

1992
Edmond H. Fischer (1920–) and Edwin G. Krebs (1918–) 'for their discoveries concerning reversible protein phosphorylation as a biological regulatory mechanism'.

1993
Richard J. Roberts (1943–) and Phillip A. Sharp (1944–) 'for their discoveries of split genes'.

1994
Alfred G. Gilman (1941–) and Martin Rodbell (1925–98) 'for their discovery of G-proteins and the role of these proteins in signal transduction in cells'.

1995
Edward B. Lewis (1918–), Christiane Nüsslein-Volhard (1942–), and Eric F. Wieschaus (1947–) 'for their discoveries concerning the genetic control of early embryonic development'.

1996
Peter C. Doherty (1940–) and Rolf M. Zinkernagel (1944–) 'for their discoveries concerning the specificity of the cell mediated immune defense'.

1997
Stanley B. Prusiner (1942–) 'for his discovery of prions—a new biological principle of infection'.

1998
Robert F. Furchgott (1916–), Louis J. Ignarro (1941–), and Ferid Murad (1936–) 'for their discoveries concerning nitric oxide as a signalling molecule in the cardiovascular system'.

1999

Günter Blobel (1936–) 'for the discovery that proteins have intrinsic signals that govern their transport and localization in the cell'.

2000

Arvid Carlsson (1923–), Paul Greengard (1925–), and Eric R. Kandel (1929–) 'for their discoveries concerning signal transduction in the nervous system'.

2001

Leland H. Hartwell (1939–), R. Timothy Hunt (1943–), and Paul M. Nurse (1949–) 'for their discoveries of key regulators of the cell cycle.'

2002

Sydney Brenner (1927–), H. Robert Horvitz (1947–), and John E. Sulston (1942–) 'for their discoveries concerning genetic regulation of organ development and programmed cell death'.

Name index

NOTE: **Bold** figures refer to entries in the list of Nobel Laureates in the Sciences, 1901–2002 (pp. 305–334).

A

Abraham, Edward 151
Adrian, Edgar Douglas **327**
Ahlquist, Raymond 200
Alder, Kurt **319**
Alferov, Zhores I. **315**
Alfvén, Hannes **311**
Altman, Sidney 99, 132, 179, 180, 196, **323**
Alvarez, Luis W. 137, 176, **311**
Amaldi, Edoardo 157, 227
Anderson, Carl David **308**
Anderson, Philip W. 7, 166, 217, **312**
Anfinsen, Christian B. 140, **321**
Anne, Princess 215
Appleton, Edward Victor 135, 136, **309**
Arber, Werner **332**
Archimedes 118
Arrhenius, Svante August 18, 77, **316**
Astbury, William 53, 54, 89, 132, 170
Aston, Francis William 136, **317**
Avery, Oswald T. 58, 59, 81, 89, 141, 142, 159, 169, 221, 225, 226
Axelrod, Julius **331**

B

Baekeland, L. H. 190
Baltimore, David 142, 215, 218, **331**
Banga, Ilona 207
Banting, Frederick Grant 7, 67, 221, **326**
Bárány, Anders 13, 51, 235, 240
Bárány, Robert 202, **326**
Bardeen, John 69, 101, 144, 165, 175, **312**
Barkla, Charles Glover **306**
Bartholomew, J. R. 33, 238
Bartlett, Neil 229, 230
Bartlett, Paul D. 143
Barton, Derek H. R. 71, 89, 91, 107, 208, 239, **321**
Basov, Nicolay Gennadiyevich 45, **311**
Bastiansen, Otto 249
Beadle, George Wells 130, 159, **329**
Becquerel, Antoine Henri 76, **305**

Bednorz, J. Georg 80, **313**
Beethoven, Ludwig van 85
Bell, Jocelyn 51, 236, 239–41
Belousov, Boris 193
Benacerraf, Baruj **332**
Benson, Andrew A. 231
Benveniste, Jacques 28
Berg, Paul 70, 88, 113, 118, 121–2, 152, 163, 175, 176, 213, **322**
Bergius, Friedrich 71, **318**
Bergmann, Max 141
Bergson, H. L. 127
Bergström, Ingmar 233, 235–6
Bergström, Sune K. 74, 106, 123–4, 161, 215, **332**
Beria, L. P. 115
Bernal, J. Desmond 53–4, 63, 129, 154–5, 163, 222, 223, 249
Bernstein, Richard 155–6
Berry, Stephen 86
Berson, Solomon 196, 241
Berzelius, J. J. 46
Best, Charles 7, 67, 221
Bethe, Hans Albrecht 52, 157, 158, **311**
Binnig, Gerd **313**
Biot, Jean Baptiste 100
Birge, Raymond 239
Bischitz, Lajos 36
Bishop, Michael J. 76, 216, **333**
Black, James W. 35, 63, 73, 122–3, 130, 144, 200, 208, **333**
Blackett, Patrick Maynard Stuart 63, 136, **309**
Blobel, Günter x, 7, 70, 109, 124, 141, 161, 196, **334**
Bloch, Felix **309**
Bloch, Konrad 143, 161, **330**
Bloembergen, Nicolaas 151, **313**
Blumberg, Baruch S. **331**
Bohr, Aage 76, 130, **312**
Bohr, Niels 36, 44, 76, 136, 138, 191, 232, 234, 235, **307**
Boltzmann, Ludwig 233

Index

R

S